D0272246

AGRICULTURAL PROCESSING FOR DEVELOPMENT

Agricultural Processing for Development

Enterprise Management in Agricultural and Fish Processing

JOHN C. ABBOTT
Former Chief, Marketing and Credit Service, Food and Agriculture Organization of the United Nations

Avebury

Aldershot · Brookfield USA · Hong Kong · Singapore · Sydney

Published by

Avebury

Gower Publishing Company Limited,
Gower House, Croft Road, Aldershot,
Hants, GU11 3HR, England

Gower Publishing Company,
Old Post Road, Brookfield, Vermont 05036
USA

British Library Cataloguing in Publication Data

Abbott, J.C. (John Cave), *1919*-
Agricultural processing for development.
1. Developing countries. Food processing industries. Management
I. Title
664'.0068

ISBN 0 566 05553 8

Printed and bound in Great Britain by
Athenaeum Press Limited, Newcastle upon Tyne

Contents

Tables

Preface

This book is the first on its subject, the management of agricultural and fish processing enterprises in the developing countries. It deals with rice, maize and cassava milling, the preparation from fruits and vegetables of canned products, juices and preserves, the drying of flowers and herbs, cotton ginning, the processing of sugar cane, tea and tobacco, of livestock and their products, and of fish.

The book is also new in its approach. It begins by presenting 40 case studies of processing enterprises that have succeeded. They are drawn from a range of African, Asian and Latin American countries. The cases are then used to illustrate kinds of enterprise suited to particular tasks and sets of conditions. Private indigenous, transnational joint venture, cooperative and parastatal enterprises are assessed in this context.

Processors have been under fire in some quarters for taking food production resources away from lower income consumers, putting at hazard the long run productivity of significant land areas and neglecting the environment. These issues are examined from an objective stance. Agricultural and fish processing enterprises are making major contributions to economic and human development. This is being recorded locally and nationally. These benefits will be the greater and more extensive when they are backed by appropriate government policies and support.

Agricultural processing projects conceived in government development plans or promoted by bilateral aid have been characterized by a high rate of failure. A substantial proportion of the 'white elephants' in the developing world are officially sponsored processing plants. They have failed because the potential supply of raw material for processing has been over estimated, the marketing function has been neglected, or convenience in management has been sacrificed for potential economies of scale that were never realized. These problems are examined. Pragmatic approaches to their solution are presented. Management of the processing enterprise as a business operation is given special attention.

This book is designed for two main sets of readers. It will be immediately useful to those responsible for agriculture and fisheries development in third world countries and in the aid agencies concerned. Its practical tone will also appeal to those directly involved in the establishment and operation of processing enterprises. It is also designed for use as a supplementary text in agricultural economics and marketing at the universities of the developing countries and in courses offered elsewhere for students from those countries. Discussion topics and exercises are presented. They focus on processing as a business in students' own countries and on the factors affecting its success.

Prices and financial accounts have been converted into US dollars at the official rate of exchange at the time quoted.

Acknowledgements

The cases Haji Mansur, Enebor, Nigeria and Chitali's Dairy were taken from 'Private enterprise and rural development', FAO, 1985 by M. Harper and R. Kavura. Chalam's Herbochem and Jamhuri Tannery were also kindly provided by Professor M. Harper, Cranfield Institute of Management. 'Frozen orange juice concentrate', and 'Lobster tails Brazil' were adapted respectively from papers by R.J. Strohl and D. Levy in 'Successful agribusiness management', Gower, 1985 edited by J. Freiwalds. He also wrote 'Jamaica Broilers' for 'Agribusiness Worldwide' October/November 1983 from which this case is taken. 'Ratna Feed Industries, Nepal' and 'Pineapple canning, Siam Food Products' are based on papers respectively by P.R. Mathema and C.Y. Lee included in 'Marketing - an accelerator for small farmer development', FAO, Bangkok, 1976. 'Cassava pelleting in Thailand' is based on a paper by Boonjit Titanivatanakan 'Transnationals in the development of cassava exports from Thailand', Faculty of Economics and Business Administration, Kasetsaart University, Bangkok. 'Charoen Pokphan group' is taken from 'Food processing and marketing in Thailand' UNCTAD, Geneva

1985 by N. Poapongsakorn et al. 'Taiwan Kagoma Food Company' is derived from 'Farmer factory contract system for processing tomato' by M.R. Meneguay and K.R. Huang, Asian Vegetable Research and Development Centre, Taiwan 1976. The Corn Product Corporation International experience was presented by A. Schumacher, World Bank at a symposium held at the University of California, Berkeley in 1975. 'Peppers for Tabasco' is taken from a case study by G.A. Truitt originally prepared for the Harvard Business School, and 'Botswana Meat Corporation' from one by H. Mettrick, International Course for Development Oriented Agriculture, Wageningen. 'Adams International, Thailand' and 'Mhlume Sugar Co., Swaziland' are based on studies by Business International Corporation for AID, Washington in 1985. 'Rose's lime products, Ghana' and 'Dabaaga Fruit and Vegetable Canning Co.' were developed from information provided respectively by E. Reusse and E. Seidler, FAO marketing specialists. 'Swaziland Meat Corporation' is based on a report by Hunting Technical Services 1983 and 'The broiler boom in Lebanon' on one by K. Reda, FAO, 1970. 'Frozen shrimps Ecuador' is derived from 'ITC roving seminar on the processing and export marketing of shrimps', ITC, Geneva 1985 and 'Fishmeal Peru' from 'Fishing for growth' by M. Roemer, Harvard University Press 1970. 'Allana frozen fish, Bombay' and 'Nouadhibou, Mauritania' have been adapted from papers by R. Baynes, and E. Hempel and V. Kwong, which appeared in 'Info fish marketing digest' 2, 1984 and 5, 1985 respectively. The balance of the case studies were prepared by the author from information available through FAO or obtained by direct contacts and enquiries.

Grateful appreciation is expressed of the assistance received from FAO and, in particular, from H.J. Mittendorf, Chief Marketing and Credit Service. The author also wishes to thank the United Nations Conference on Trade and Development (UNCTAD) for permission to present case materials originally prepared for its use, and the Cambridge University Press for permission to reproduce sections of cases already featured in 'Agricultural marketing enterprises for the developing world', 1987.

Abbreviations

AID	Agency for International Development, Government of the USA, Washington
CDC	Commonwealth Development Corporation, London
C.i.f.	Cost, insurance, freight i.e. the buyer pays for a product delivered to an agreed destination
COMECON	Council for Mutual Economic Assistance, an organisation for economic cooperation between various states in the socialist world
COPAC	Committee for Promotion of Agricultural Cooperatives, c/o FAO, Rome
EEC	European Economic Community
FAO	Food and Agricultural Organization of the United Nations, Rome
F.o.b.	Free on board i.e. the buyer is responsible for the costs and risks of transport

IDA	International Development Association, World Bank, Washington
ITC	International Trade Centre, Geneva
ODA	Overseas Development Administration, Government of the UK, London
OECD	Organisation for Economic Cooperation and Development, Paris
UNCTAD	United Nations Conference on Trade and Development, Geneva
UNCTC	United Nations Centre on Transnational Corporations, c/o UN, New York
UNIDO	United Nations Organisation for Industrial Development, Vienna
WFP	World Food Programme, Rome

1 Introduction

Farm and fisheries products are processed in order to preserve them over substantial periods of time and make them more convenient in use. The treatments applied range through drying, pickling, freezing and canning fruits and vegetables, husking paddy, grinding maize, the slaughter of livestock, gutting of fish and preparation of the meat for consumption, to the more complex processing of oilseeds into margarine and cane into sugar. Essentially, there is a change of form as distinct from the cleaning, sorting and refrigeration of fresh produce. Such processing can extend greatly the sales life of perishables, present products in forms more attractive to consumers, facilitate transport over long distances and overcome a wide range of marketing constraints.

Many of the processes involved are carried out most effectively with the aid of specialized equipment in plants assuring the necessary sanitary and other conditions. They involve major fixed investments. To cover the capital and other overhead costs the operator needs a steady supply of suitable material for processing and a dependable marketing channel for the processed product. To meet the technical requirements of a particular process and the preferences of consumers, the material for processing may have to be chosen specifically. It must also be available at a price that permits operation at a profit taking into account market returns and the costs incurred. For these reasons the supplying of a processing enterprise needs careful planning and organisation.

If the processing operation is successful it can offer a stable and profitable outlet for the farmers and fishermen who supply it. It can add variety to popular diets and make food available at times when otherwise it would be lacking. Exports of processed products can be an important source of foreign exchange. Potentially, therefore, the processing enterprise is a powerful engine of development.

VALUE ADDED IN PROCESSING

For some agricultural and fisheries products processing is essential if they are to reach consumers in an acceptable form. Many others benefit from processing through extension of the range and duration of the markets in which they can be sold. In this section we shall review briefly the contribution processing can make in increasing the value of farm and fishery output.

Rice is harvested as paddy with a protective husk. This has to be removed before cooking. Most consumers like white polished rice so the second layer of bran is also removed. Bran can be used directly as a livestock feed ingredient, or first processed to obtain bran oil. The husk can be burned to provide steam to power the mill. There is a strong preference for whole grain. Keeping the grain intact during milling is important. Milling efficiency is commonly judged by the percentage of whole grain obtained from paddy. In the Indian sub-continent and West Africa paddy is often soaked in water and steamed before milling. This process known as parboiling gelatinizes and thus hardens the grain leading to higher milling yields. The efficiency of alternative milling techniques is summarized in Table 1.1.

Table 1.1 Product yields from paddy by milling technology

	Husk	Bran	Husk and bran	Whole rice	Broken rice	Total rice
	(.			percent. . . . .)		
Hand pounding	-	-	40	40	20	60
Steel hullers	-	-	36.6	46.5	16.9	63.4
Disc shellers	-	-	32.5	55.9	11.6	67.5
Rubber rollers	22	8	30	62	8	70

Source: Esmay et al. (1979)

Maize is ground into grits, coarse meal and flour for use as human food and for inclusion in livestock and poultry feed. In much of Africa south of the Sahara and in parts of Latin America white maize is a staple food. Local mills have replaced hand pounding at the village level. Large scale mills have been set up to serve city populations. A wet milling process has been developed to yield edible oil, sweeteners and starch; interest in maize base sweeteners was promoted by the high international sugar price of 1974: their use has continued to expand. A 120 ton per day modern maize mill more than doubles the output of a 50 ton per day mill at very little more cost. Economies of scale are still greater in wet milling.

Cassava has long been processed for direct food use in the countries where it is grown. It is peeled, grated and squeezed to remove the acid element, then roasted. Gari processing in Africa includes a stage of fermentation. Cassava is also processed industrially for export as tapioca and flour for alcohol distillation, sago, starch and glue preparation. The great development of the 1960s was the processing of cassava into pellets for export to Europe as an ingredient in livestock feed. The raw roots are chipped, air dried, passed through a hammer mill, then pressed into cylindrical shaped pellets. These are convenient to handle and reduce greatly the shipping space required. This procedure was developed with the discovery of a rich market for cassava in Europe via a loophole in the EEC system of protective import controls.

Feed mixing is one of the fastest growing agricultural industries. This is related to the increasing human demand for animal protein. Nutritionally balanced feeds are particularly important for the most efficient grain to protein converters - poultry and pigs. They are also widely used for cattle and sheep to supplement natural grazing and in weaning young animals. A wide range of grain, oilseed and other product combinations and supplements can be drawn on to meet desired nutritional requirements at the lowest cost. Reliability and consistent uniformity are essential from the nutritional point of view and because the ingredients of mixed feeds cannot be assessed visually by the buyer. Ironically, while animal feed mixing now focusses specifically on nutritional and protein content with high levels of quality control, in most processing for human use nutrition is a minor concern.

Processing is strategic in expanding the market for perishable fruits and vegetables which mature during a limited season. Drying, pickling with brine and vinegar, jam making, canning and preservation of juice under low temperatures open the way to additional markets in time and space, and in consumer appeal and convenience. Internationally, the demand for canned products has levelled off since the 1960s reflecting the expansion of frozen foods in home consumption and of dried vegetables for industrial use. However, the demand for canned exotic fruit and juices such as papaya, mango and passion fruit is increasing.

Wild and cultivated herbs with medical applications are a significant resource of many rural areas. They are the major ingredient in most traditional treatments for human ailments and those of domestic animals. Contemporary preferences for 'natural products' favour wider use of such herbs in the medical and health maintenance programmes of developed societies.

Sugar cane is crushed to expel the juice which is then boiled to evaporate off the water. The end product is brown sugar known locally as jaggery or gur. Quantities of up to 15 tons per day are commonly handled in this way. Centrifugal processing to produce a white sugar is generally based on a throughput of 1,000 tons per day. Refining to obtain white crystal sugar offers substantial economies of scale with 100,000 tons of product annually considered the minimum.

Tea is marketed after the leaves have been allowed to wilt and ferment, and have then been chopped and dried. Critical factors in the processing of good quality tea are that only the bud and first two leaves are picked and that the leaves be brought to the factory for processing the same day. They can be transported easily, but the plant must still be relatively near to the production area. Picking and processing can continue through the year. One hundred kg. of fresh green leaf produces 18 to 27 kg. of tea. Economies of scale in processing call for a production base of 200 to 400 ha. Packing tea in consumer sized packages adds about 25 percent to its value, in tea bags 100 percent. However the cost of importing the packaging reduces the net profit. Individual processing enterprises may also have difficulties in offering blends adapted to discriminating consumers' requirements.

Tobacco leaves are cured by drying on racks, generally with the use of artificial heat. In Jamaica, Kenya and Nigeria this is undertaken by growers on a family basis, or in groups. They operate under contract to a tobacco company which provides seedlings and other inputs on credit, plus intensive technical assistance. In Thailand tobacco is grown by small farmers under contract, but grading and curing is centralized.

Cotton fibre or lint is harvested together with the seed. For effective marketing they must be separated at an early stage. This process, known as ginning, permits the fibre to be packed tight in bales for sale to buyers for spinning. The seeds go to crushing plants for manufacture into oil and livestock feed. Because cotton as harvested is so bulky, it is ginned near to the point of production. Where cotton is picked by hand, mature clean seed cotton can be collected separately from immature, damaged or stained growth and kept free of leaves and stalk. Cotton picked by machine is more variable in quality and requires extensive cleaning.

Edible oils for use in cooking are obtained from olives, cotton seed, groundnut, oil palm, sesame, soya, sunflower and other oil seeds. The oil is expelled by crushing and is then refined. The refined oil can then be sold directly or further processed into margarine and other solid fats. The residue is an animal feed with a high protein content. Traditionally oil has been expelled by grinding and pressing on an artisanal scale. With such methods the residual cake still contains about 10 percent oil. Use of a mechanical press can reduce this to 5 percent. Solvent extraction can bring the oil level down to 1.5 percent: it is a more complex process involving higher capital investment. Mechanical presses are generally used for capacities of up to 200 tons of raw material per day and for seeds with high oil content. Vegetable oils are refined to achieve a product that is light in colour and bland in flavour.

In particular, the proportion of free fatty acid should be below three percent. However, in some developing countries consumers are accustomed to much higher proportions and prefer oils with the more pronounced flavours that go with this.

The slaughtering of livestock and preparation for consumption of the carcass and offal is a specialized operation in most societies. Chilling, freezing and canning facilitate marketing over long distances. Preparation and packing of special cuts add further value where adapted to market requirements.

Cow hides and goat and sheep skins are a valuable by-product of livestock production. They deteriorate rapidly unless they are processed. Adhering flesh must be cleaned off and the skin treated with preservatives.

In the past meat poultry were killed, plucked and dressed for individual consumers on request. With access to refrigeration, the development of fast growing strains and integrated production and processing, the young chicken has shifted from a luxury eaten on special occasions to the cheapest source of meat. Average yields from live weight of product for retailing to consumers are beef 44, veal 50, lamb 45, pork 65, chicken 70 and turkey 78 percent respectively.

Much milk is consumed near where it is produced without any processing. However, transporting milk to urban customers involves cooling and pasteurization if its quality is to be assured. Conversion of milk into products such as butter, cheese and evaporated milk extends greatly the market. Drying into powder facilitates storage over time. The powder form is also convenient for manufacturing uses, for reconstitution into liquid milk and for blending with water and high fat content fresh milk. Processing into casein opens up another outlet for milk surplus to other uses.

Fish are highly perishable. Consumption in the developing countries has been limited to a few kilometres distance from where they were caught.

Traditionally they have been preserved by salting, drying and smoking. Canning or freezing brings them to the consumer with less change in form and opens the way to a much wider range of markets. Until 1960 shrimp caught by fishermen in India were discarded, or spread around coconut trees as fertilizer. Exported frozen, they earned $45 million in foreign exchange in 1985 from the USA alone. Meal derived from drying and grinding fish that could not otherwise be sold, is a valuable high protein ingredient for livestock and poultry feeds. Fishermen trawling for shrimp in South East Asia often pick up three tons of 'trash' fish with every ton of shrimp. Much of it was thrown back into the sea. In 1980-82 27 percent of the world's fish catch was converted into meal.

SCOPE FOR NEW PROCESSING INITIATIVES

A massive expansion of agricultural and fish processing in the developing world is foreseen in official projections such as that made by FAO for the year 2000. A considerable growth in processed exports from developing to the more developed countries is foreseen. Still more important will be the expansion needed to serve exploding city populations in the developing countries themselves. These are the conclusions of projections of population growth, food consumption and national development requirements.

Exports

These will grow both in response to market opportunities because of lower costs in the developing world, and under government pressure to earn foreign exchange. At any one time, the export market for processed agricultural products may seem very limited. Access to many of the richest markets may seem blocked by quotas and duties designed to protect higher cost domestic suppliers. Markets for traditional tropical exports such as coffee, tea, and palm oil may seem saturated. Nevertheless, opportunities keep opening up. Dutch and German animal feed mixers found a way to import annually six million tons of cassava pellets from Thailand. Negotiation at the government level resulted in quotas of processed products that could be imported into EEC countries being assigned to many African countries under the Lomé Convention. On this basis an asparagus canning enterprise was started in Lesotho that could never have been envisaged otherwise. Political interest led the USA to open up formerly protected markets to 24 countries of the Caribbean.

It follows, therefore, that those interested in exporting processed agricultural products should look systematically for opportunities to:

a) replace or supplement existing suppliers on markets that are open to them;
b) enter a market that is currently protected for which an import quota or more favourable entry terms might be negotiated at the political level.

A third approach in opening up new export markets is via barter agreements or counter trade. Thus large quantities of frozen orange juice might be exported from Brazil, for example, against imports of petroleum products. Counter trading opens up further possibilities. The exporter of orange juice agrees that he will accept in return various products or money equivalent in total value to that established for the orange juice. These may come to him via a third country. Traditionally, this has been a way of developing exports to the COMECON countries of Eastern Europe. Counter trade is also increasing outside the centrally planned countries. OECD has estimated that five percent of the trade between other countries is on such a basis. Growing centres for these arrangements are Miami, Singapore and Vienna. Disadvantages of barter trade are the need to take products in return that may be over priced or only partly satisfactory, together with the complications, costs and uncertainties involved.

Overall, it is still worthwhile for a potential exporter of processed products to investigate thoroughly the options open to him.

Domestic markets

The starting point for many processing enterprises in developing countries has been import substitution. They set out to replace a fairly widely consumed import with a product based mainly on local raw materials. Governments have fostered this to save foreign exchange.

The opportunity to substitute a local product for imports is the most tempting of domestic market opportunities. With protection against import competition the project may seem secure. However, if protection results in the price charged to domestic consumers for the processed product being substantially higher than it was before, the market could melt away. This can lead into a vicious circle whereby less is sold, costs rise again because of the weight of overheads, and the domestic market contracts still further. A policy conclusion for government is - let some imports continue. Their quality and price will set a target for the domestic enterprise to match. The need to face this competition will also serve in good stead as a defence against government or other interference that might result in increased costs.

Much more important for the longer run are the new markets forthcoming from population increase, rising incomes and, above all, people moving into cities. Over the next two decades the populations of the developing countries will continue to grow rapidly, in spite of the relative success being achieved with programmes for family planning. Strategic, for the marketing of agro-industry products is the number of urban residents dependent on the market for their food supplies, and their ability to pay for processed

Table 1.2 City population projections; developing countries, 1990 and 2000

	1980	1990	2000
	(. . . millions . . .)		
Amman, Jordan	0.7	1.0	1.5
Bangkok, Thailand	4.7	7.0	10.0
Blantyre, Malawi	0.4	0.9	1.5
Bombay, India	8.3	11.8	16.8
Cairo, Egypt	7.4	9.9	12.9
Calcutta, India	8.8	11.7	16.4
Colombo, Sri Lanka	4.0	5.8	8.1
Dhakka, Bangladesh	3.0	6.0	10.5
Dar es Salaam, Tanzania	1.1	2.5	4.6
Jakarta, Indonesia	7.2	11.0	15.7
Karachi, Pakistan	5.0	7.8	11.6
Kathmandu, Nepal	0.2	0.3	0.5
Kuala Lampur, Malaysia	1.1	1.7	2.4
Lusaka, Zambia	0.7	1.4	2.3
Maseru, Lesotho	-	0.1	0.2
Nairobi, Kenya	1.3	2.8	5.3
New Delhi, India	5.4	8.0	11.5
Seoul, Rep. of Korea	8.4	11.5	13.6

Source: Department of Economic and Social Affairs, UN, New York, 1980.

foods. City populations in developing countries are expected to grow by 2.5 to 4 percent a year in all regions. Thus by 2000, the urban population of developing market economy countries is projected to be no less than 2,200 million, double that of 1980. Cities whose present populations will treble by the year 2000 include Dhakka, Dar es Salaam, Lusaka and Nairobi. (See Table 1.2) Bangkok, the three largest Indian cities, Cairo, Djakarta, Karachi and Seoul will all have populations over 10 million. These are the obvious domestic growth markets for processed agricultural and fish products.

Table 1.3 Expected increases in crop processing: 90 developing countries, 1980-2000

	Quantity	Annual growth rate
	tons millions	percent
Wheat	84	3.0
Rice	135	2.6
Coarse grains	59	2.4
Vegetable oil	63	4.3
Sugar	67	3.8
Fruit and vegetables	10	6.4
Cotton ginning	19	4.0
Total	437	3.1

Source: FAO, (1981) Agriculture: toward 2000, Rome

Income growth and urbanization in the developing countries will call for an increasing share of total output to be processed and to be processed to a higher degree. Under the twin impact of growing urban population and income, the volume of agricultural produce processed through the primary stage, i.e. milling of grain, slaughtering of meat animals, is estimated to double between 1980 and 2000. (See Table 1.3) Processing of products highly responsive to income changes and important in urban as compared with rural diets, such as sugar, vegetable oils, fruit and vegetables, will rise by about 5 percent per annum. Processing of basic staples such as cereals will expand more slowly, by about 3 percent per annum, although processing for feed use will rise much more rapidly.

FAILURES AND SUCCESSES

To obtain a realistic base for advising on projects to set up food industries in developing countries, FAO undertook in the 1960s a study of some 70 plants that had not lived up to expectations. All of these plants appeared to be well designed from an engineering point of view. The factors responsible for losses or failure were of a different nature. In several cases, there had been mistakes in planning; either market demand for the product or raw material supply had been over-estimated. With others poor management resulted in excessive operating costs, rendering them unable to compete with other enterprises supplying the same market. The incidence of such difficulties is shown in Table 1.4. The groupings are as follows:

1. Problems of raw material supply, including over estimation of potential supply, lack of suitable varieties for processing, insufficient incentives to farmer suppliers and lack of production support services such as extension and credit.

2. Problems of market demand, including over-estimation of prospective demand; misjudgement of tastes, preferences and habits of consumers; under estimation of competition from other sources and substitutes, and of obstacles to entering foreign markets.
3. Problems of management, in particular poor handling of raw material, product storage and distribution; lack of marketing management and sales promotion; inefficient internal management including mobilisation of working capital, collection of payments due, and overstaffing; inappropriate government interference.

Table 1.4 Incidence of difficulties encountered by processing enterprises

	Number facing major difficulties		
	Raw material supply	Market demand for product	Management
Slaughterhouse	13	6	19
Dairy	8	4	5
Fruit and vegetables	24	11	2
Rice	6	-	-
Oilseeds	2	-	3

Source: Mittendorf, H.J. (1968) 'Marketing aspects in planning agricultural processing enterprises in developing countries', FAO Monthly bulletin of agricultural economics and statistics 17 (4).

For some projects, no proper market outlet studies had been conducted before a decision on the investment had been taken. In others it was simply assumed - without investigation of production response - that the establishment of a processing plant would attract a supply of raw material. Such errors in decision making still occur. Of seven government sponsored cassava processing plants built subsequently in Venezuela only two were still operating in 1980.

In Chapter 2 of this book we present short case studies of 40 processing enterprises in the developing countries that have been notably successful. Some such as the Anand dairy cooperative, the Kenya Tea Development Authority, Jamaica Broilers and the Botswana Meat Corporation have attracted wide international attention. Others smaller and less well known are representative of many such enterprises that have also contributed greatly to the well being of farmers and fishermen, and of consumers, as well as to the economies of their country.

The products for which processor case studies are presented are: rice, maize, cassava, livestock feed, fruits, vegetables, flowers, medicinal herbs, sugar, tea, tobacco, cotton, oilseeds, beef, hides and skins, poultry, milk, fish and fish meal. These cases also provide a wide geographical coverage of the developing world. Additionally they reflect both the processing of raw materials from traditional patterns of agriculture, livestock raising and fishing, and of material produced specifically for processing under a new fully integrated system. Cases such as the broiler boom in Lebanon and fish meal in Peru have been overtaken by events. They have been included to illustrate intermediate stages of producer/processing/marketing integration that may be relevant to conditions in some areas.

The relative suitability of alternative forms of processing enterprise for various purposes and sets of conditions is taken up in Chapter 3. This is discussed in terms of materials to be processed, capacity to organize a regular flow of raw material supplies from farmers and fisheries, and ability to market processed products effectively to export and domestic customers.

The contribution that the processing enterprise can make to development and how governments can be of assistance is the theme of Chapter 4. Specific attention is given to the transmission of technology and to welfare effects on both agricultural and fishery producers and consumers, and on the rural population as a whole.

The final Chapter 5 deals with the planning and management of a processing enterprise in the developing countries. Its role is to systematize, and supplement where necessary, what can be learned from the practical experience presented in the case studies.

ISSUES FOR DISCUSSION

1. Construct profiles of agricultural processing enterprises in your own country or a defined part of it drawing on published data, press cuttings, annual reports where issued and information obtained by visits and interviews.

2. Do these enterprises provide an adequate coverage for the agricultural and fish production of the area treated?
   What additional enterprises do you consider to be needed?
   On what scale would they operate?

3. Have there been in your country significant projects to establish agricultural or fisheries processing plants that have failed?
   If so, what were the reasons?

FURTHER READING

The following list includes, in addition to material on the economic and business aspects of agricultural and fish processing, a selection of authoritative texts on processing technology. They are provided for readers who also want a technical coverage.

Akehurst, B.C., (1981) 2nd edit., Tobacco, London, Longman Group.

Balkow, V.E., (1982) Manufacture and refining of raw cane sugar, Amsterdam, Elsevier Scientific Pub. Co.

Blackburn, F., (1984) Sugar cane, London, Longman Group.

Bruinsma, D.H., W.W. Witsen and W. Wurdemann, (1983) Selection of technology for food processing in developing countries, Wageningen, Pudoc.

Cook, J.H., (1985) Cassava: new potential for a neglected crop, Boulder, Westview Press.

Considine, D.M., ed. (1982) Foods and food production encyclopedia, New York, Van Nostrand Reinhold Co.

Courtenay, P.P., (1980) Plantation agriculture, Boulder Co., Westview Press.

Cullison, A.E., (1979) 2nd edit. Feeds and feeding, Reston, Reston Publishing Co.

Das, R., (1981) Appropriate technologies in cereal milling and fruit processing industries, New York, Vantage Press.

Esmay, M., E. Soemangat and A. Phillips, (1979) Rice post production technology in the tropics, Honolulu University Press of Hawaii.

FAO, Agricultural bulletins and development papers: (1962) Animal by-products processing and utilization; (1970) Cashew nut processing (1977) Cassava processing; (1983) Maintenance systems for the dairy plant; (1962) Dates: handling, processing and packaging; (1972) Fruit juice processing; (1985) Hides and skins improvement in developing countries; (1984) The retting of jute; (1978) Packaging, storage and distribution of processed milk; (1975) Olive oil technology; (1974) Rice milling equipment, operation and maintenance; (1984) Rice parboiling; (1973) Processing of natural rubber; (1978) Slaughterhouse and slaughter slab design and construction; (1980) Small scale cane sugar processing and residue utilization; (1974) Tea processing.

Fisheries technical papers and circulars: (1977) The production of dried fish; (1972) Equipment and methods for improved smoking and drying of fish in the tropics; (1985) Planning and engineering data: fish canning; (1984) Planning and engineering data: fish freezing; (1985) The production of fish meal and oil.
Marketing Guides: (1977) 2nd edit. Marketing livestock and meat; (1976) 2nd edit. Marketing fruit and vegetables; (1972) Rice marketing; (1961) Marketing eggs and poultry, Rome, FAO.

Gerrard, F., (1977) 5th edit. Meat technology, London, Northwood Publications.

Grist, D.H.,(1986) 6th edit. Rice, London, Longman Group.

Hanson, L.P., (1975) Commercial processing of vegetables and (1976) Commercial processing of fruits, Park Ridge, New Jersey, Noyes Data Corp.

Harben, P. and H. Cooke eds., (1985) A practical guide to counter trade, New York, Metal Bulletin Inc.

ILO/FAO (1982) Small scale processing of fish, Geneva, ILO.

ILO/UNIDO (1984) Small scale maize milling, Geneva, ILO.

Kent, N.L., (1983) 3rd edit. Technology of cereals, Oxford, Pergamon Press.

Killick, T., (1978) Development economics in action - a study of economic policies in Ghana, New York, St. Martin's Press.

Kohel, R.J. and C.F. Lewis eds., (1984) Cotton, Madison, Wisconsin, American Society of Agronomy.

Koll, H.A.J., (1987) The economics of oil palm, Wageningen, Pudoc.

Pfost, H.B. and Pickering D., eds. (1976) Feed manufacturing technology, Arlington, American Feed Manufacturers' Association.

Tropical Development and Research Institute: An introduction to fish handling and processing; A guide to the economics of dehydration of vegetables in developing countries; Slaughter facilities for tropical conditions: a guide to the selection and costing of appropriate systems; Quality control in the animal feed stuffs manufacturing industry; An industrial profile of cotton ginning; Monographs on the processing of particular tropical products and markets for the processed product and by-products.

UNIDO: Guides to information series - (1975) Canning industry; (1977) Coffee, cocoa, tea and spices; (1976) Dairy product manufacturing industry; (1974) Leather and leather processing industry; (1976) Meat processing industry; (1972) Vegetable oil processing industry.

Weiss, E.A.,(1983) Oilseed crops; London, Longman Group.

2 Case studies of processing enterprises

HAJI MANSUR'S RICE MILL, INDONESIA

This mill was started in 1951 in Sidoarjo, a large town about 25 kms. from Surabaya, capital of East Java. It was situated on family owned property. Paddy was bought from farmers nearby; it was milled by hand on a beaten earth floor, using pounders and sticks. Every 100 kg. of paddy produced 50 kg. of rice and 50 kg. of husk and bran. The rice was sold to wholesalers on the local market. Mansur continued the business in this way until the early 1960s. He earned a reasonable profit by buying in times of glut and selling when rice was scarce.

In the 1960s the government expanded its rice supply and price stabilisation programme. This was implemented through a parastatal Bureau of Logistics (BULOG). Mansur saw an opportunity to operate as a buying agent for BULOG. He obtained a contract to supply it with rice at pre-announced prices. With this in hand he was able to borrow $80,000 to mechanize his operation. Storage was constructed and trucks acquired to bring in paddy and transport rice to BULOG's depot near Surabaya and to other outlets.

Table 2.1 Income and expenditure: Haji Mansur's rice mill, 1979

	$	$
Income		
Sales to wholesalers	288 000	
Sales to BULOG	1 152 000	
Total		1 440 000
Expenditure		
Purchase of paddy	992 000	
Milling costs	16 000	
Interest charges	24 000	
Depreciation	40 000	
Sales, administrative and other expenses	120 000	
Total		1 192 000
Profit		248 000
Taxes		119 040
Profit after taxes		128 960

In 1979 the annual turnover was $1,440,000 with 15 full time employees. They handled about 4,000 tons of rice a year. The bulk of this was sold to BULOG with a small proportion going at higher prices to wholesalers. Tables 2.1 and 2.2 summarize the financial position of the business.

As buying agent for BULOG the mill paid 25 cents per kg. for paddy and received 36 cents per kg. for milled rice: it retained the bran. This margin provided a continuing incentive for efficiency via: maintaining the mill in condition to maximize the out turn of rice; operating the mill to capacity; minimizing intermediary costs in obtaining supplies.

To this end Haji Mansur advanced funds to farmers to help meet their production costs so assuring a substantial direct supply. Sales to independent wholesalers of milled rice selected to meet special quality requirements brought a higher margin, but the quantity was relatively small.

Table 2.2 Financial statement: Haji Mansur's rice mill at 31 December 1979

Assets	$	Liabilities	$
Cash	15 000	Bank overdraft	100 000
Accounts receivable	45 000	Accounts payable	55 000
Stocks	63 000	Accrued liabilities	25 000
Advance to farmers	75 000	Long term loan	80 000
Equipment and vehicles net of depreciation	192 000	Family investment and retained earnings	130 000
Total	390 000	Total	390 000

In obtaining supplies Haji Mansur was in direct competition with farmers' cooperatives also acting as agents for BULOG. They had the advantages of access to low interest loans for their equipment and government-provided transport to BULOG depots. Haji Mansur was able to compete satisfactorily on the basis of his personal contacts with farmers and ability to advance them credit when needed. A long term disadvantage, however, was the increasing industrialisation of Sidoarjo. A number of the farmers who used to supply him with paddy had sold their land for industrial development; he had to draw rice from an ever widening area, thus significantly increasing his transport costs. It would be desirable to relocate the mill in a main production area. Whether such an investment was justified when it would receive none of the subsidies and special support available to his co-operative competitors was a major question mark.

HANAPI AND SONS, MALAYSIA

This mill was established in 1980 by a family owned company registered as Hanapi and Sons Private Ltd. Its authorised capital was $130,000 of which $90,000 was paid up in 1984. It was located in "the rice bowl" of Malaysia, 11 km. from Alor Setar, capital of Kedah state. Its physical structures and equipment comprised a paddy drying platform, stores, mill building and office, huller, polisher, mechanical dryer, weighing and moisture testing equipment and diesel power units. Capacity per hour was 3 tons of polished rice. Normal working hours were 12 per day for 26 days per month. Paddy drying capacity was 1,768 tons per month designed to match its milling capacity. The average out-turn was polished rice 65, brand and polishings 8, very fine rice 3, broken rice 1.5, and husk 22.5 percent respectively.

Supplies of paddy were obtained mainly from farmers, wholesalers and the National Paddy Board in the nearby Tajar district. Farmers delivered directly to the mill and were paid cash, likewise the wholesalers. Total paddy production in Tajar was around 11,000 tons. The mill's requirements were 8,600 tons, but it faced competition from four other large scale mills located within 8 km. distance. It believed that paying farmers cash gave it an edge over its competitors. However, supplies were also obtained via wholesalers from more distant producing areas.

Marketing was mainly under a continuing contract to supply the Felda Trading Corporation of Shah Alam, Selangor. The price followed that announced by the National Paddy Board for grade B1 rice, i.e. $36 per

100 kg. Rice was also sold to local wholesalers at the B2 grade price of $33 per 100 kg. Obtaining the Felda contract improved considerably the profitability of the operation. Returns on sales rose from 14 percent in 1982 to 31 percent in 1984 and on investment from 16 percent to 44.5 percent. The by products were also sold to wholesalers.

Credit for 45 days was afforded the Felda Trading Corporation, and local wholesalers 15 days. By-products were sold for cash. On average the mill waited 35 days for payment. One month would be more usual.

Management was compact and efficient. Zakaria Hanapi, the largest shareholder was managing director. The operations division was headed by his brother. Mr. Zakaria's wife was in charge of administration and accounts. All the 10 workers employed in the mill were on continuing contracts with payment according to the amount of work done. They received free lodgings and meals. The Managing Director, Mr. Zakaria Hanapi had been running the mill since its inception in 1980. He had attended various courses in paddy business. His brother had been buying paddy in nearby areas for over five years. He was a paddy farmer himself. Mrs. Zakaria had been with the mill for over five years. In 1986 she had attended an accounting course specific to paddy operations organised by the National Paddy Board.

For construction of the mill and initial operating funds the company obtained a term loan of $100,000 from the Agricultural Bank in 1979. By 1985 this had been cleared off. Further credits of $40,000 in 1983 and guarantees for overdrafts of $90,000 and $330,000 were obtained. These were repaid and a new loan of $400,000 to buy paddy from farmers was taken in 1984.

All these loans and credit facilities were backed by physical collateral such as agricultural land and debentures on fixed and floating assets. In addition, the Board of Directors stood jointly and severally as guarantors. Favoured by its indigenous Malaysian status the Hanapi enterprise received substantial

Table 2.3 Income and expenditure: Hanapi and Sons, 1982 and 1983

	1982	1983
	$	$
Income		
Sales: rice	200,000	265,000
bran, polishings	21,800	17,000
very fine rice	6,200	3,000
Milling charges	9,200	-
Transport charges	7,000	-
Total income	244,200	285,000
Less purchase of paddy	145,000	150,000
Gross margin	99,200	135,000
Expenditure		
Transport	11,500	10,700
Vehicles	6,000	-
Wages, salaries, allowances	1,300	5,600
Equipment, spare parts, repairs	6,000	2,100
Diesel oil	3,800	3,800
Sacks	6,000	2,600
Telephone, electricity	600	900
Insurance	2,700	3,600
Bank interest	4,200	2,700
Depreciation	9,500	13,000
Donations	400	-
Miscellaneous	700	800
Total expenses	52,700	34,800
Net profit	46,500	100,200

assistance from the Bank. The Bank also saw advantages to the farmers in the area around. They were able to sell their paddy directly to the mill and be paid in cash.

Income and expenditure figures for the mill are presented in Table 2.3. The Bank considered the mill's liquidity position to be strong with current ratios at 2.2 (1982) and 1.6 (1983), indicating ability to settle current liabilities out of its immediate current assets. It had attained a successful operation by virtue of its competent management, sound business strategies and timely bank financing. The benefit to paddy farmers in the area was substantial since it constituted new well financed competition for the existing structure of millers and wholesalers.

ENEBOR, NIGERIA

Enebor had a good job in Benin, the capital of Bendel State in Nigeria, but he disliked city living. In 1961 he responded to a government sponsored "Farmers' Crusade" and acquired a two hectare rice farm in the Illushi Plains. This was considered an area of high potential. Every September the flood waters of the Niger backed up its tributaries covering the plains and depositing rich alluvial silt. This provided an ideal environment for rice growing. New varieties had been introduced, and appropriate fertilizing, cultivation and other practices developed. Energetic farming entrepreneurs were needed to take advantage of the opportunities it offered.

Initially Enebor continued to live in Benin travelling to Illushi to supervise the farming operation. As the need for labour increased, the able bodied members of his family moved to Illushi permanently. Early in the 1970s, Enebor finally gave up his job and moved his whole family to Illushi. He started a small trading enterprise to supplement his income. With savings and through hire purchase arrangements he acquired rice husking and milling equipment worth several thousand dollars. He became one of the most successful rice farmers and processors in the area. His farm grew from two to 22 hectares; his total enterprise had an annual turnover exceeding $30,000.

A simplified operating statement for Enebor's enterprise is presented as Table 3.4. Much the most important source of income was parboiling and milling paddy as a service for other farmers. This brought in $17,600 during the year of the study. The rice grown on his own farm cost $4,840 in tractor hire, labour, fertilizer, rent and other costs. It seems unlikely that it would be profitable if the cost of milling was not covered by the income from custom milling for others. Enebor complained of high cost labour and difficulties obtaining seed and fertilizer. He considered it inadvisable to expand further in farming. It would be less risky to devote his efforts and funds to building up his profitable business of contract milling and sale of rice grown by others.

Illushi itself had only a population of 1,000. The sales outlet for rice after milling was its regular weekly market which attracted traders from the north and the rest of Bendel State as well as from Imo, Anambra and River states. The overall demand for rice in Nigeria was large and increasing. In part it was met by imports, but the demand for locally produced rice was not satisfied. Enebor and other growers and processors succeeded in developing the reputation of "Illushi rice" as a preferred quality to the point that paddy from other areas was channeled there for processing.

In other respects Illushi was not an advantageous location. It had no piped water or electricity. Mills had to be run on petrol or diesel oil, both very expensive. There was no hospital or dispensary and only one primary school. Because there were so few services very few people wanted to live there. This compounded the labour problem. There was no good road to the nearest main centre 70 km. away. Though the market was well attended it was very poorly equipped. A proposal to establish a cooperative that would be eligible for concessional credit

Table 2.4 Income and expenditure account: Enebor

	$	$
Income		
Charges for parboiling, milling and related services		17,600
Sales of rice	13,760	
less cost of own production	4,840	
purchases from other farmers	8,280	
	640	640
Total income		18,240
Expenditure		
Wages - manager and skilled operator	7,680	
Wages - casual labour	2,780	
Fuel	1,090	
Repairs	450	
Depreciation	800	
Rent of premises	480	
Total expenditure	13,280	13,280
Net income		4,960

and attract government support for improvements in infrastructure had not materialized.

Overall, Enebor had a profitable business. It constituted an outlet for rice growers locally and further afield. It provided employment for a number of people. To expand further and contribute more effectively to the rich productive potential of the area, it awaited improvements in basic infrastructure.

CORN PRODUCTS CORPORATION INTERNATIONAL, KENYA AND PAKISTAN

In various developing countries Corn Products Corporation the American multinational has sought opportunities to apply its wet processing technology. End products are starch, sweeteners, oil and gluten feed and flour for livestock. On average some 40 percent of the wet starch is dried and 60 percent converted to sweeteners. For its operations in Kenya and Pakistan Corn Products Corporation formed companies with local investors in which it took a 51 percent share.

The Kenya operation was located at Eldoret in the northwest surplus maize producing area. Throughput was 7,000 to 10,000 tons of maize per year. Used machinery was brought from Corn Products Corporation plants in other countries. Originally procurement was via an eight year purchasing contract with the Kenya Cereals Board, which had a monopoly of wholesale purchases from farmers. When the Board sought to raise the price sharply, the plant eventually obtained authorisation to buy directly from farmers. Contracts were made with large growers including one public sector company. The plant employed 150 persons of whom three were expatriates.

In Pakistan Corn Products Corporation bought into Rafhan Maize Products Co. the largest maize miller in the country. It was located at Faisalabad in the Punjab where the main food grain was wheat. Capacity was expanded and new finishing channels added. It was the first company in Pakistan successfully to manufacture maize sweeteners. Of the 95 percent yield which Rafhan achieved on the maize it ground, 67 percent was converted into starch, three percent was oil and 25 percent remained as a component for cattle and poultry feed; some of the starch was further processed into glucose, dextrose, and syrups.

Maize was procured through commission agents buying in rural assembly markets. Normally they allowed buyers for domestic food use to take up their requirements first. Purchases for Rafhan were then made at a 'reserve' price announced in the newspapers and on the radio. Farmers could also sell directly at the mill, but few did so. The supplies obtained were never sufficient for capacity operation even though the company provided hybrid seed to the government services, and allowed credit to farmers to build storage cribs. Finally, Rafhan undertook on its own account a programme to develop improved seed, an extra (spring) crop, and to provide inputs on credit together with extension to farmers linked to it by individual contracts. (See Appendix 1) The average size of farm contracted in 1975 was 16 has. Deliveries to the mill of corn on the cob were made directly by farmers or in transport arranged by Rafhan extension staff (6 agronomists). In 1975 Rafhan obtained 20,000 tons from contract farmers divided more evenly between summer and spring crops so reducing the pressure on its working capital to finance stocks in storage. The yields obtained by the farmers under contract were twice the national average.

Rafhan has also contributed to the economy of Pakistan through import substitution and by generating foreign exchange. In normal years 6,000 tons of starch were exported to the Middle East. Its investment in plant, machinery, and equipment since acquisition by Corn Products Corporation amounted to over $5 million. About $1 million of this was financed by USAID and Pakistan Industrial Credit and Investment Corporation loans. Internally generated profit and cash provided the balance. Rafhan has also stimultaed expanded production of maize for sale on the cash market as well as through its contracts.

CASSAVA PELLETING IN THAILAND

The area planted to cassava in Thailand increased from 38,400 has. in 1957 to 960,000 has. in 1977. Output rose from 0.4 million tons in 1957 to 12.4 million. Exports at $528 million earned more than any other commodity. Credit for this development goes to the European feed importers. They saw the opening offered by the EEC system of protective duties. These raised the price of locally grown feed grains far above the levels prevailing in exporting countries outside the EEC. Cassava was not subject to such duties because it was not considered a major source of feed supply for the Common Market. Also, there were no producers inside the Community with immediate interests to be protected. Acceptance of cassava as a carbohydrate ingredient in animal feed mixes was facilitated by the increasing sophistication of the compounding industry. The deficiency in protein could be made up by an additional input of soy meal.

The Germans invented the cassava chipper and established in Thailand a German made hammer mill to make cassava meal. Some of the early shipments to Europe suffered from mould growth. A. Toepfer applied to cassava the process of pelleting flour with three percent of molasses. In 1967-68 German investors put $1.0 million into the first pelletizing plant in Thailand.

Preeminent in organising market flows were Toepfer, Krohn and Co. and Peter Cremer. These firms traded in other feed ingredients for the European market. They had the necessary market expertise. They could mobilize finance. They were in a position to charter whole vessels and so cut transport costs. Several had integrated horizontally or vertically. Krohn did both, by expanding its own pelleting capacity and also by forming in 1976 a group of satellite suppliers. Later Tradax (Cargill) came in, investing over $20 million in port facilities for pellets in Amsterdam. In 1981 it built the largest hard pellet plant in Thailand.

Growers sold directly to some 1,500 cassava chipping plants which generally arranged for transport from the farm. This must take place immediately after harvest; fresh roots should not be held for more than four to six days. After sun drying the chips were sold to pellet factories. In 1982 there were 385 pelleting factories scattered throughout the production areas. Most had their own trucks for transporting their input (chips) and output (pellets). Many pellet factories were set up by rice millers, and jute and kenaf processors.

Financing began with a letter of credit to the Thai exporter opened by the transnational's European office. The exporter pledged this at the Bank of Thailand obtaining 90 percent of its value at an interest rate of 7 percent, with an obligation to export the specified shipment within 180 days. This low interest rate was established by the government to promote agricultural exports. On the basis of such a credit, or its expectation, funds went back through the rural millers to the growers. Millers competing for supplies to match their capacity made contracts with farmers. The millers supplied the inputs needed and purchased the entire crop. Competition between them often secured the same price for farmers accepting credit as for those selling on the open market. The cost incurred by millers in advancing such credit was covered by the saving on overhead costs with mills operating near to capacity.

Cassava growers did very well during this period. A study of marketing margins in 1982 showed them receiving for unpeeled roots 37 percent of the price of the equivalent quantity of pellets c.i.f. Rotterdam. Of the price received at the farm one third was net margin after accounting for all costs. In addition to a secure market the growers enjoyed the production advantages of easy and cheap propagation, relatively high yields, low costs of production, and good risk aversion. A prospective cost for the longer run is soil exhaustion and erosion. In the meantime cassava pelleting for export has made a major contribution to rural incomes in some of the poorer parts of Thailand and to the overall development of the country.

RATNA FEED INDUSTRIES, NEPAL

Ratna Feed Industries belonged to a family that had various business ventures and close government contacts. The feed company was established in 1966 with a loan from the Nepal Industrial Development Corporation. It started with the following interrelated objectives:

1. To manufacture and sell feeds to meet the increasing demand;
2. To promote poultry and livestock farming; and
3. To provide technical know-how, veterinary facilities, and market outlets for poultry, and for poultry and other livestock products.

It took on the poultry development role because it recognized the potential market and saw the need for coordination of breeding, management, veterinary and credit services. Sister enterprises were set up to produce chicks and provide veterinary services. Eggs were handled by a branch of the feed firm.

In the mid seventies the firm supplied over 90 percent of the specialized poultry feed used in the Kathmandu Valley. Dealers received feed on credit; they were paid a commission of five percent. Where they could achieve a quick turn over the dealers' income from feed handling was high. This brought them into promoting poultry farming. In 1974/75 Ratna distributed more than half of the broiler chicks in the Valley. The marketing and technical services unit of Ratna Feed provided technical and economic know how to the new poultry raisers. It also helped solve problems of management, feeding, and housing. Ratna issued pamphlets on poultry diseases and their control, poultry management and feeding, with other useful information. These were especially helpful to the smaller producers.

The Veterinary Medical Store (PVT) Ltd., a sister company of Ratna, maintained stocks of drugs and vaccines needed by poultry raisers. Veterinarians were available at the central office in Wotu Tole. This office also maintained a team of vaccinators with their own means of transport.

Scattered throughout the Valley, the branches and dealers of Ratna Feed Industries also served as outlets for the eggs of poultry keepers. Ratna tried to maintain a steady flow of eggs on to the market to minimize price fluctuations. Operating on a relatively large scale for Nepal it was able to buy and sell eggs for a margin of 1.5 to 3.8 cents per egg in 1968/69. This brought down the margin taken by other dealers, and poultry raisers received an additional 3 to 4.6 cents per egg. The establishment of an additional channel for marketing eggs raised the level of competition and brought this benefit to producers and consumers.

Ratna Feed did not handle meat poultry, possibly because of religious susceptibilities. However, it encouraged some local meat dealers to handle them, providing refrigerators on credit to these retailers. Ratna also furnished technical and marketing know how, and advertised to promote the sale of broilers.

In summary, the initiative and organising capacity of a feed processing firm and the timely provision of new techno-economic know how was a major contribution to the growth of the poultry industry in Nepal. Its role in organising the marketing of poultry products helped to keep down prices to consumers, and so stimulated domestic demand.

PINEAPPLE CANNING, SIAM FOOD PRODUCTS

Siam Food Products was established as a pineapple canning company in 1970. Its plant was built in the Chon Buri area with the intention of drawing on a different production base from that of two already established competitors - Dole Thailand and Thailand Pineapple Cannery. For the farmers in Chon Buri sugarcane and cassava were the traditional crops. Pineapple was grown on a relatively small scale for the fresh market.

To serve Siam Food Products' markets in Europe, Japan and the USA, 50,000 tons of field pineapple were required. It planned to grow one-third of these on land it had purchased and to obtain the other two-thirds from independent farmers by contract. The soil was suitable. The necessary inputs and technology would be furnished by the company. It approached the Chon Buri Farmers' Association which welcomed the idea and circulated a notice to its members. In total, 133 farmers with land ranging from two to 150 has. applied for a grower contract. Contracts with these 133 farmers were signed in early 1971.

According to the contract, SFP would provide the entire package of production inputs. It would buy pineapple suckers and supply them to the farmer, also fertilizers and herbicides. The farmer would also require labourers to prepare the land, grow and harvest pineapples. As it would take 18 months for pineapple to be harvested, this would entail a substantial investment. SFP therefore met the labour costs in cash. It also provided technical assistance. In turn the farmer agreed not to use the inputs provided by SFP for any other purposes. After harvest, he would sell all the pineapples produced to SFP's cannery in Ban Bung village near Chon Buri town. SFP would pay $14 per ton of fresh pineapple delivered at the cannery. The farmer would receive $4.80 in cash and $9.20 would be deducted towards repayment of the credit.

Only a few of the farmers kept to the contract. Of the 10,000 tons expected, only 3,000 tons were actually sold to the company in 1972/73, and 674 tons in 1974/75. According to SFP's management, the following factors led to this poor performance:

1. Low yields. The production technology recommended was much more intensive than that customary in the area. Many farmers were unable to follow it.
2. Supervision was inadequate. Four company field men had to deal with 138 farmers scattered over an area 100 km. in radius.
3. The price offered was considered too low. Since the time the contract was signed the fresh market price had risen to $22 per ton.
4. A short sighted view on the part of many farmers regarding repayment of the company's credit.

In 1974 the company wrote off half the $348,000 of credit still due from the farmers. It decided to expand direct production to meet 80 percent of its requirements and buy the balance from whoever was willing to supply at the prevailing price.

The conclusions of the company's plantation superintendent were that:

a) Contract growing should have begun with a few selected farmers and expanded on the basis of experience.
b) It was impractical to fix the price in advance. Since a firm outlet was offered for large amounts it could have been set at a given margin below the prevailing fresh market price.

ROSE'S LIME PRODUCTS, GHANA

After cocoa, coffee and cola the only noteworthy agricultural based export from Ghana during the 1960s was lime products originating from Rose's plant at Asebu. In 1966 the company shipped in barrels 5 million litres of raw and filtered lime juice valued at 13 to 19 cents per litre, 13,000 kg. of lime oil worth $12 per kg., plus skins in halves and shreds. An additional plant was being installed to process skins as raw material for pectin production.

Marketing was assured by the parent company in England which produced the internationally known "Rose's Lime Juice". Eventually it became part of the transnational Cadbury Schweppes. Originally drawing its supplies from a plant in Grenada in the Caribbean, Rose's started up in Ghana as a potential source of additional supplies to meet a growing demand at lower cost.

This plant was still operating in the 1980s surviving various military take overs and long periods of unrealistic exchange rates. That this was possible is attributed to:

1. Limited domestic demand for the raw material. Lime supplies built up for factory demand over several years could not be absorbed by the local market. Therefore, their prices remained relatively undisturbed by the general upward trend of agricultural produce prices in domestic markets.
2. Shipment of a crude product in an inexpensive wholesale package, keeping the foreign exchange component in f.o.b. costs below 30 percent.
3. Confinement of fixed investments to what was essential, e.g. wooden structures as far as feasible, clean airy sheds instead of closed halls for processing operations. As a result the total investment inclusive of working capital amounted to only one third of potential annual sales at capacity operation.
4. Utilization of small farmers' efficiency supported by good extension services, instead of running a plantation directly.
5. Sober overall management and fair dealing with farmers who received a price equivalent to 50 percent of the f.o.b. value of the products derived from their supplies.

Finance for lime tree planting - a long term investment - came from development sources. Current inputs were limited. Processors' staff met farmers each season, provided management advice and sometimes made advances. They helped clear access roads. Their goal was to maintain juice output near to the plant's capacity of 9 million litres annually.

FROZEN ORANGE JUICE CONCENTRATE, BRAZIL

The growth of the frozen concentrated orange juice industry in Brazil has been remarkable. The first plant came into operation in 1962. By 1981 total annual output had reached 490,000 tons (65° Brix,

65 percent solid). By that date 140 million trees producing seven million tons of oranges had been planted and the number was still growing. In 1984 citrus juice exports brought in $1,400 million of foreign exchange.

The juice plants began as initiatives of North American processors such as Norton Simon looking for alternative, cheaper or additional sources of supply. They brought with them the latest technology including that for by-products - D limonene, essential oils, essence oils and citrus pulp pellets. Cargill pioneered shipment by bulk tankers. It also handled a large share of the citrus pulp. Another large processor Citrisuco was part owned by a drinks manufacturer in Germany. Many of the smaller plants had links with the four market leaders and together accounted for over 90 percent of frozen juice exports.

While several of the big processors had large groves under direct management the bulk of the fruit was grown by independent farmers. It was a business fraught with risk. Export prices for juice have fluctuated widely in response to frosts in Florida which have been the great stimulus to development in Brazil, and surplus conditions when the Florida crop was ample. The frost of Christmas 1983 brought the price of oranges to growers in Brazil up from $0.85 to $3.00 per box.

While processors and fruit growers within convenient transport distance were dependent on each other, negotiation of a mutually acceptable price has often been protracted. On occasion producers organised strikes to withhold deliveries. When the Sanderson plant went bankrupt it was taken over as a producers' cooperative Fruitesp. Eventually the government set a minimum export price, maintained by a licensing system (CASEX), for which foreign exchange must be repatriated. It also required that a uniform price to producers be negotiated annually. This still left room for differences over fruit quality, time of payment, etc.

Seedling nurseries were mostly private. However the federal and state (Sao Paolo) governments have supported research, pest control and extension. Canker control has been financed by a levy of 1.4 cents per box delivered. Both processors and producers have received generous government credit at interest rates well below that of inflation. There have also been tax concessions for processors and export subsidies in some years. In the face of increasingly frequent surpluses the government has cancelled these promotional measures. A programme of storing frozen concentrated juice to help stabilise the market has been under consideration.

DABAAGA FRUIT AND VEGETABLE CANNING CO., TANZANIA

A fruit and vegetable processing enterprise that has operated successfully in Tanzania for more than ten years is the Dabaaga plant at Iringa. This was established and run by a man of Goan background who was himself a farmer. He obtained a loan from the Rural Development Bank to purchase the plant which came from India as a turnkey operation. An Indian food technologist advised on its installation and operation.

The plant's output of jams, pickles, juices and soups was diversified to keep it going through the year and to match the requirements of a relatively low income domestic market. Its products were marketed through local food distributors under an attractive brand label "Life".

With low overhead costs this plant was designed to run on the basis of ongoing farm production of tomatoes, peaches, plums, etc. from the operator's own and neighbouring farms.

Seeds and fertilizers were purchased through normal commercial channels. While the processor had an account with the National Commercial Bank and could obtain an overdraft there, small scale growers of fruit and vegetables in Tanzania were unlikely to qualify for institutional credit. There would, however, be a government horticultural extension specialist for this area. While the operator provided informal advice and assistance he had no specific contracts with producers to secure supplies. The plant was run on the basis of the operator's own production and supplies available nearby that were seasonally in surplus on the fresh market.

TAIWAN KAGOMA FOOD COMPANY

Processed tomato products are in growing demand for food use in most developing countries. The Taiwan Kagoma Food Company undertook the canning of tomato products primarily for export. However, the procedures followed in developing the necessary supply are directly applicable for plants serving domestic markets, and have been followed widely.

The Taiwan Kagoma Food Company (TKFC) was formed in 1967 as a joint venture between Taiwan merchants and Japan's Kagoma Company. Its primary objective was to produce canned tomato products, of which 60 percent would be exported to Japan. The processing plant was located in Shanhua, Tainan County, because the climate and existing cropping systems were especially suited for a full winter crop such as tomato. It was also thought that the farmers there could be trained to grow processing tomatoes. The main concern was to obtain a cheap supply of processing tomatoes reliable in quality, quantity, and time of delivery.

The resources of the factory management for forming a tomato supply system were: (i) previous experience in Japan of organising farmers to produce efficiently and deliver tomatoes to a factory; (ii) improved varieties (subsequently tested at their factory in Taiwan) and cultural practices developed by their Japanese parent company's research unit; (iii) technical extension staff who were acquainted with farmer extension techniques; (iv) a factory located within a potential tomato production zone; and (v) an industrious farming community that could shift crops within their cropping systems if they were assured of a price incentive.

Extension staff were recruited from local cooperative and government service personnel. Their job was to select and supervise farmer representatives, coordinate the distribution of inputs to the farmer representatives, and provide technical assistance, such as improved cultivation methods, insect and disease control measures. Groups of farmers located in suitable areas were contracted to put some 20 has. under tomatoes. For each group a farmers' representative was selected on the basis of farming, leadership and bookkeeping ability. His responsibility was to (i) coordinate team production and harvesting; (ii) raise the company's improved varieties and distribute the seedlings; (iii) store and distribute inputs, such as pesticides and fertilizer, for the company; (iv) maintain records of company farmer transactions; (v) to advise and supervise pest control campaigns; and (vi) to distribute the company's wooden crates at harvest time. For these services, he earned the money from selling the seedlings and 0.26 cents per kg. for all tomatoes harvested and brought to the factory by his team.

The contract offered to farmers provided for:

1. Provision of seedlings (at a price of 0.13 to 0.18 cents each) - ensuring uniform quality and harvesting time for a given location.
2. Provision of pesticides and fertilizer on credit with repayments handled by the farmer representative.
3. An agreed price per grade delivered e.g. 3.2 cents per kg. first grade and 2.4 cents second grade in 1974/75. An additional 0.13 cents per kg. was paid for early varieties and 0.08 cents for later varieties as an incentive to permit extension of the processing season from December to May.
4. Harvesting dates set by the company and field crates provided by it to keep supplies in line with plant capacity.

Three shifts were worked per day during the peak season.

5. Farmer delivery of supplies to the plant by ox cart or motor vehicle with payment by the company according to the distance.
6. Grading by random sampling of each delivery with grade specifications defined rigorously and agreed mutually.
7. Farmer commitment to deliver his total output under penalty of no further contracts. This was supervised by the farmers' representatives.
8. Countersignature of each contract by two operators who were responsible in case of non repayment of production loans.

Payments to growers were made two to three weeks after delivery. Advantages of this system to the company were (i) direct control over the production and procurement of processing tomatoes; and (ii) ability to increase or decrease quickly production locations according to market conditions. A sample of growers interviewed in 1975 saw the following advantages to them:

a) guaranteed purchase of their whole crop at a pre-determined price;
b) provision of healthy seedlings and technical guidance;
c) the crop brought higher returns than other winter crops and required only a short growing season.

Dissatisfaction was voiced over:

a) delays in provision of crates because of factory over capacity: this resulted in tomatoes rotting in farmers' fields;
b) lower grades received because of delays in handling farmers' produce after it had arrived at the plant.

Since the farmers' representative was effectively a company man some arrangement for independent mediation was needed. With the issue of grower/processor bargaining power coming up also with asparagus and mushrooms, mediation over prices, etc. by a Taiwan government department became accepted practice.

PEPPERS FOR TABASCO, HONDURAS

Traditionally, peppers for McIlhenny's Tabasco sauce were grown in Louisiana near its processing plant. By the late 1960s, with demand increasing, it faced supply problems over its raw material. Labour for harvesting was scarce: mechanisation proved impracticable. Scouting for alternative sources of supply McIlhenny made contracts with a group of farmers in Honduras. Later the farmers found that they could not live up to the contracts.

Assistance was proposed by a religious organisation La Fragua which had received funds for the support of agricultural projects following the hurricane Fifi. It proposed a contract between the McIlhenny Company and La Fragua which would then arrange with farmers to produce hot peppers. This was accepted; the company advanced some funds, and a small processing plant to inspect, clean, trim, mash and salt peppers in wooden barrels was built on the property of La Fragua. In 1977, the project was working with 55 small groups of farmers. La Fragua passed on to these groups seed and instructions on cultivation supplied by McIlhenny.

In 1980, 86 groups produced 342,000 kg. receiving 56 cents per kg. La Fragua made a profit of $17,810. For 1981, contracts were signed with 90 groups representing 300 farmers with prices ranging from 56 to 66 cents per kg. At this stage finance was obtained from the National Agricultural Development Bank and the original credit from McIlhenny was repaid. These results were obtained by a combination of efforts. The farmers provided their land and labour. The

The Centre conducted field seminars followed by demonstrations and provision of advisory materials. The company provided the technology via an agronomist stationed at La Fragua, and visits by other experts three or four times a year. Most important of all, it provided an assured market at a fixed price growing at about 10 percent annually. By requiring that a quarter of the value of peppers' deliveries be retained in a savings fund, La Fragua set about building up a reserve to finance future activities.

Once the project was well underway, the company financed a second plant in the Valley of Lean. It covered the operating costs of both processing plants including salaries of the 13 employees. It also paid 10 percent of the cost of grinding to La Fragua in compensation of its expenses plus $2.50 per barrel of processed peppers for storage.

The project had advantages for each of the parties concerned. McIlhenny obtained the supplies it needed with little investment, exposure to risk, or commitment by management personnel. It made good use of local initiative, development agencies, and transport facilities. The Honduran farmers had a long term contract that brought them $1.0 million annually. This had two advantages as against selling on the spot market i.e. less price volatility, and the possibility of resolving pricing disputes in the context of an ongoing relationship. The project made good use of intermediation assistance as provided by La Fragua. Its technicians were able to communicate the company's needs and standards to individual producers. This helped solve quality control problems, without costing the company money.

In 1985 La Fragua had ceased to be involved. However, McIlhenny was still drawing supplies from the primary processing plants in Honduras.

BASOTHO FRUIT AND VEGETABLE CANNERS

Asparagus is well suited to the soil and climatic conditions of Lesotho. It grows there wild, but until recently it had never been grown commercially and it still does not appeal to local consumers. However, the canning plant set up at Mazenod a few years ago to take advantage of EEC import preferences provided under the Lomé Convention has proved successful. It has earned valuable foreign exchange and operated at a profit.

The project received a capital contribution from the UN Capital Development Fund and technical assistance from FAO. It then continued on its own as an operation of the Lesotho Development Corporation. It marketed four grades of asparagus with a higher proportion of green stalk and purple heads than competing supplies. Together with a 17 to 20 percent duty preference on entry to the European Common Market this has secured it an advantageous outlet in Germany. It began by skimming the top of the market, selling to the highest bidder: later it shifted to contracts with several EEC importers.

Asparagus was a new product in Lesotho. Together with the government extension service the project had to develop production of its raw material from zero. It set up its own nursery. It supplied farmers, mostly women, with clones, prepared the soil for them, and provided sprays and fertilizers for a crop which takes four or five years to mature. It undertook to buy all the output of the 240 farmers under contract in 1986 at a price about 2 cents per kg. above that prevailing over the border in South Africa. Packing containers and trimming equipment were provided. Collectors made their rounds daily. The plant also financed the cost of boreholes to provide water for supplementary irrigation and for washing the plants after harvesting. Shade huts to keep stalks cool and moist while awaiting collection have also been provided.

Problems seen by the plant manager in 1986 were those of persuading farmers to maintain their access roads in passable condition, the 13 percent financing charge to be met once it had to be obtained from a commercial bank, and the limited operating season. The asparagus crop provided only three to four months work. Trials were being made with beans and peas as complementary lines.

This project demonstrates a relatively passive approach to marketing with high assumption by the processor of responsibilities for raw material production. Extension assistance and financing guarantees were provided by the government.

MARIGOLD HEADS, ECUADOR

In the 1970s marigold became an important crop for small farmers in the Sierra region of Ecuador. It was particularly well suited to families with only a few hectares of land. It was a crop in which they had a production advantage because the flowers had to be picked at a certain stage of maturity. Picking the heads when they reached this stage induced the plant to produce more. Going over a field of marigolds in this way a number of times during a season could be undertaken by family labour at low cost.

The flower heads are dried and processed to produce a colorant used in balanced feeds for flocks of layers kept under intensive battery conditions. It promotes a stronger yellow in the egg yolks.

The initiative for the introduction of this crop came from a private firm which set up a processing plant in Quito and exported to the USA. It distributed seed to farmers on land suited to the crop, instructed them on management, advanced credit, and undertook to buy the dried flower heads. Agents were employed to collect them on regular visits to the contracted producers.

Initiative, technology and finance and a sure market were provided by the processing firm. Its contacts with farmers were direct through its own agents.

CHALAM'S HERBOCHEM, INDIA

This enterprise processed a range of herbs, meat and other rural products including rice polishings and orange peel into extracts for use in traditional and modern drugs. It began with oleo resins and belladonna gradually building up to 40 different products.

The founder and owner Mr. Chalam took a degree in chemistry followed by a diploma in business management. He then worked in a company that produced herbal medicines. From 1960-71 he was production superintendent of Biochemical and Synthetic Drugs Ltd. With this experience he started his own company. He obtained physical premises on a hire purchase basis from the Ardra Pradesh Small Scale Industry Development Corporation and a loan of $20,000 from the State Bank of Hyderabad under a Technocrat scheme.

Herbochem employed fourteen people; six skilled workers, four unskilled, two chemists, an office assistant and a part time accountant. Their monthly wages ranged from $20 to $80 with a regular bonus, free uniforms and insurance. The factory worked in two shifts, from 6.00 a.m. to 2.00 p.m. and then from 2.00 p.m. to 10.00 p.m. The processes used were percolation, or heating and filtering, with solvents. The percolation method took around four to six days. The raw material was powdered and mixed with the solvent. The following morning the liquid was collected and distilled. When alcohol was used the process was repeated three to four times until all the soluble extracts had been removed. When water was used as a solvent distillation was accelerated by putting the boiling kettle under pressure. A variety of chemicals were used to assist the cooking and filtering process and preservatives were added. The company was equipped to handle two processes at the same time typically running each on a batch basis for five to six days.

Herbs and other supplies were obtained from assemblers in various parts of India and from a specialized firm in Bombay which had its own assembling agents. Many of the herbs were available only in certain seasons. Advance deposits were made to assure the firm of continuing supplies. Major customers included Ralli and Glaxo. Most sales were made directly without any intervening agent. The company also exported to five customers in Hong Kong. Customers were normally allowed two to three months credit;

Table 2.5 Income and expenditure statement: Chalam's Herbochem, year ending 31 March 1982

	$
Income	
Sales of processed products	54,870
Closing stocks	20,679
Total	75,549
Less	
purchase of raw materials	29,113
opening stocks	17,361
Gross operating margin	29,075
Expenditure	
Wages, etc.	3,956
Electricity	233
Coal	5,411
Solvents, etc.	822
Transport	741
Excise duty	1,100
Total	12,263
Net operating income (before interest depreciation and taxes)	16,812

some took longer to pay. Severe price fluctuations were characteristic of this business because of the seasonal nature of raw material supplies and periodic large orders of product users unrelated to periodicity of supply.

Income and expenditure accounts and a balance sheet for 1982 are presented as Tables 2.5 and 2.6. The

Table 2.6 Balance sheet: Chalam's Herbochem, year ending 31 March 1982

	$
Assets	
Building, equipment, etc., at valuation	35,074
Owner's equity allowing for loss of $282 carried forward	5,569
Closing stocks	20,679
Debtors	18,070
Deposits	1,720
Cash on hand	206
Total	81,318
Liabilities	
Bank advances	46,716
Loans outstanding	14,151
Other creditors	20,451
Total	81,318

balance sheet carries forward a loss of $282 from the previous year. Chalam had difficulties in obtaining alcohol which was subject to allocation by a state government department. He was obliged to use peptone in its place. This was extracted from meat; suitable grades were available only at high cost. However, Chalam was confident of solving the problem of his alcohol supply and envisaged opening a second plant to manufacture food preservatives as soon as he could obtain the necessary financing.

MHLUME SUGAR CO., SWAZILAND

This company was formed in 1958 as a joint venture between the Commonwealth Development Corporation (40 percent) and Sir John Halett and Sons (60 percent). It acquired 5,200 has. of uncleared land. By 1961 it had a sugar estate of 3,000 has. processing the cane into 40,000 tons of sugar. The sugar company was established in conjunction with a Commonwealth Development Corporation (CDC) project to develop a partially irrigated area on the Mozambique border. Eventually CDC bought out the private partners. More than $40 million was invested in irrigation, infrastructure and expansion of the processing facilities. The area has been transformed into what has been described as "a truly magnificent production system which has been profitable to all concerned and is a fine example of the productive power of entrepreneurial vision, risk capital which is ventured with a long range view and a policy of patience, and committed competent management."

Starting in 1963, thirty to forty plots per year were cleared and provided to Swazi farmers until 263 such farms had been formed. These farms, mostly four hectare lots, are on long term leases which regulate crop production, agricultural methods, construction, grazing, and so on. Sugar cane production occupies about 70 percent of the land, leaving scope for food and other cash crops.

The CDC organises the distribution of irrigation water, the operation of a tractor and equipment pools, the distribution of seed cane, fertilizer and other inputs, and the cutting and transport of cane. The cost of all of these services are at least partly recovered from farmers through user fees. The government provides four field advisors, each responsible for around 65 growers.

Deliveries of cane to the mill are scheduled by the management. Overall it receives 35 percent of its supplies from its own estate, 35 percent from CDC cultivation, 12 to 15 percent from the smallholders and 15 to 18 percent from commercial farms by contract.

Table 2.7 Income and expenditure accounts: Mhlume Sugar Co., 1977, 1979, 1981/82

	1977	1979	1981/82
	(. . . $ millions . . .)		
Sales of sugar and molasses	18.6	28.6	40.5
Less: cost of cane	12.8	20.7	26.9
mill operation costs	3.4	5.1	8.5
Mill profit	2.4	2.8	5.1
Estates profit	0.1	2.6	1.7
Other receipts	1.1	0.9	0.1
Total gross profit	3.6	6.3	6.9
Less: interest	-	1.4	2.3
income tax	1.2	-	1.3
Net profit	2.4	4.9	3.3
Dividend paid	0.4	1.6	1.2
Retained profit and appropriation to reserves	2.0	3.3	2.1

Marketing of refined sugar was originally via the South African Sugar Association. In 1964 Swaziland Sugar Association was established to control all sales on domestic and export markets. Favourable prices were obtained through the Commonwealth Sugar Agreement. When Britain entered the European Common Market Swaziland received a quota of 120,000 tons. However, there were two other mills making a total capacity of 380,000 to 400,000 tons leaving a substantial quantity for sale on local and imported external markets.

Income and expenditure accounts for Mhlume Sugar Co. for 1977, 1979 and 1981/82 are summarized in Table 2.7. In 1979 its income tax obligation was balanced out by tax allowances on mill expansion outlays.

The benefits to the country were great. The project was a continuing earner of foreign exchange amounting to $45 million in 1982. It provided gainful employment to 5,000 people. The net incomes of the smallholders growing sugar on contract ranged from $5,000 to $8,000 in 1982: they had also food crops.

The Mhlume Sugar Co. has now been taken over by a public company in which the Swazi nation has a major share. It is still managed by CDC on contract.

KUMPHAWAPI SUGAR CO., THAILAND

This Japanese/Thai joint venture was established in 1963 in accordance with the Thai Government's policy of developing sugar as a source of foreign exchange. It was also intended to help raise levels of living in the relatively poor North Eastern part of the country.

The company was started with an authorised capital of $1.4 million of which $1.0 million was paid up. The major partner was Mitsui, one of the largest Japanese trading companies. Some 28 percent of the joint company stock was held by employees, farmers, sales agents and other Thai people. To enhance local interest in the business, the company planned to expand their participation in its equity to at least 50 percent. Finance for factory construction infrastructure such as roads and bridges was obtained from Japanese Overseas Cooperation funds.

An old fashioned crushing mill was purchased at Kumphawapi, close to the Laotian border, about 550 km. from Bangkok. Its cane crushing capacity was 500 tons per day. The crucial problem was to secure adequate cane for full capacity operation. In the past, only

35,000 tons of cane had been processed annually. The new company was entirely dependent upon the farmers in the province; it was not allowed to own a plantation.

The company introduced new strains of cane and intensive use of fertilizer instead of the old field burning method of maintaining soil fertility. Loans were made for the purchase of trucks to transport increased quantities of cane to the sugar mill. Also, the farmers were guaranteed crop purchase and cash advances were made for plowing, weeding, fertilizing and harvesting.

Formerly cane was brought to the mill by ox cart over winding earth roads that were frequently washed out by rain. From 1971 to 1973 the company constructed some 200 km. of paved, all weather, eight metre wide roads and bridges. With this improved road network, the supply area has been expanded to a 40 km. radius.

In parallel with increasing cane supply, the mill's crushing capacity was expanded gradually to 5,000 tons per day; an annual production of 70,000 tons of sugar required 800,000 tons of sugar cane. The capital outlays were $17 million on plant and $12 million on infrastructure, including dust collection and waste water disposal to protect the environment. Local personnel were sent to Japan for training for middle management; others were trained locally. The goal was to expand steadily Thai participation both in running the enterprise and in its equity capital.

Marketing was assured by Mitsui an established importer into Japan of raw sugar for refining and of white sugar for distribution. It had already set up a similar enterprise in Malaysia in 1959. Since 1945, sugar consumption in Japan had been growing at a rate averaging 3 percent per year. However, it peaked at 3 million tons in 1972. Further growth was not expected to exceed that of the population i.e. about one percent per year.

The impact on the surrounding area has been substantial. The company had 500 direct employees. In 1976 it paid out $12 million to some 3,000 farmers.

It had built and repaired local roads, provided 40 deep wells for local residents, given equipment to schools and hospitals. The town of Kumphawapi had grown rapidly in population and in levels of living. It had two banks, a number of retail shops, schools and hospitals.

KENYA TEA DEVELOPMENT AUTHORITY

The Authority was established in the 1950s as an autonomous parastatal. The goal was to bring to African smallholders a valuable crop hitherto grown only by foreign owned estates. It was to:

a) be a commercial enterprise under its own direction;
b) be financed from sources other than the government;
c) have full control over smallholder tea planting, processing and marketing.

This control was considered necessary if it were to secure high tea prices, low costs and become economically and financially viable. The technical package of growing, picking, processing and marketing tea had already been tested in estate production by Brooke Bond and other transnationals. Kenya Tea Development Authority (KTDA) supplied farmers on suitable land with planting material, fertilizers, etc., and extension against reimbursement from the proceeds of leaf delivered to its factories.

In 1982 KTDA operated 27 tea factories as managing agent. Each factory was a separate financial undertaking with its own board of directors on which KTDA was represented. The intention was to build up farmer participation in the direction of the factories they supplied; 15,000 farmers held about 1.6 million shares in the 16 tea factories incorporated as public companies. This gave them an interest in KTDA performance and some voice in its operations.

Table 2.8 Income and expenditure accounts: Kenya Tea Development Authority, 1978 and 1981

	1978	1981
	(. . $ thousands	. .)
Income		
Levy on tea sales	6,680	6,426
Factory management fees	1,823	1,989
Sale of planting materials	520	576
Factory design fees	35	826
Interest	85	549
Miscellaneous	187	296
Total	9,330	10,662
Expenditure		
Planting material	372	429
Field development	1,242	2,011
Inspection and collection	2,321	3,911
Head office	1,772	4,144
Depreciation	672	1,938
Farm and training schools	(39)	(68)
Total	6,250	12,365
Surplus/(deficit) for the year	3,080	(1,703)
Exchange gain on foreign currency loans	-	564
Surplus/(deficit) transferred to development account	3,080	(1,139)

In 1980-81 KTDA sold 28 million kgs. of tea at an average price of $1.13 per kg. Growers received 18 cents per kg. for 146 million kgs. of green leaf.

The margin taken for factory, marketing and production support costs was $6 million - equivalent to 21 cents per kg. of made tea. Of the total return from sales of made tea the growers received about 80 percent.

There were two payments - the first against deliveries of leaf to collection centres. The variable annual "second payment" was geared to provide incentive for quality. It reflected the prices actually obtained for the tea from the factory at which the leaf was processed. Growers were often told to resort their bags of leaf before they were accepted for weighing. KTDA field staff constantly emphasized to growers the connection between plucking quality and second payments, and the damage to common interests if some individuals fell short. Weighing and recording procedures were kept as public as possible which also increased grower confidence in the probity of the collection staff.

Buying centres were located so that no farmer had to carry his leaf for more than 2.5 km. After inspection, the leaf was purchased and transported to the factory for processing. KTDA's policy was to purchase "two leaves and a bud", the best quality of leaf.

Marketing was by auction at Mombasa and London, and by negotiated sales to private buyers. Along with the estates, KTDA was required by the government to sell 15 percent of its output, at prices well below international levels, to Kenya Tea Packers for domestic distribution. Tea sales were made by the factories under KTDA's overall direction. Its income was derived mainly from a levy on tea sold and factory management fees, sales of planting materials, etc. See Tables 2.8 and 2.9 for sample income and expenditure accounts and balance sheets.

The Authority was run by a board of directors consisting of ministry of agriculture nominees, representatives of the Commonwealth Development Corporation its main source of capital, and of growers elected regionally. The chief executive officer was appointed by the Board. Growers' committees were used as a

Table 2.9 Balance sheet: Kenya Tea Development Authority at 30 June 1981

	$ thousands
Assets	
Fixed	
Buildings, equipment, etc.	13,162
Investments in tea factory companies	22,970
Investments in Kenya Packers	760
Grean leaf price reserve	2,197
Fertilizers, etc.	3,598
Total	42,687
Current	
Stores	464
Payments due from factories	46,134
Prepayments	8,285
Cash	7,069
Total	61,922
Less current liabilities	
Creditors	4,068
Payments due to growers for leaf	32,394
Bank overdraft	4,751
Total	41,213
Net current assets	20,709
Total assets	63,396
Liabilities	
Loan capital (CDC, World Bank, etc.)	52,870
Reserves	6,729
Mortgages	4,297
Total liabilities	63,396

channel for explaining policy, discussing proposed changes, the allocation of planting materials, etc.

The Authority had three operational branches - extension, leaf control and factories. The extension staff were seconded from the ministry of agriculture, but paid by KTDA. An assistant general manager for leaf control supervised leaf officers each in charge of 36 leaf buying clerks, plus leaf collection and transport. A chief factory superintendent supervised five group factory managers each responsible for five or six factories with their own managers, assistants and trainees.

The benefits were clear - increased foreign exchange earnings and buoyant incomes in the tea growers' region - averaging $250 annually from tea alone. Since most growers had other income, their level of living was comfortably above that of the area.

The success of KTDA began with its smallholder base. High quality plucking is difficult to maintain with hired labour. Consistently it has produced higher quality tea than the commercial estates. This has been backed up by incentives for performance built into the organisation. Grower returns were directly related to world prices realized for specific factories' output. Field staff performance was subject to grower pressure through district tea committees and in KTDA's board. Field staff and growers alike were answerable to the factories for delivering high quality leaf. Factory managements were quickly held accountable for the quality reflected in the price of the resulting made tea.

Equally important has been the stance of the government. There has been no diversion of the proceeds of tea sales and no interference in day to day operations.

ADAMS INTERNATIONAL, THAILAND

Adams International is a joint venture of a Thai-Chinese family company and the W.A. Adams Company

Inc. of Durham, North Carolina. It was formed in 1969, and began operations by exporting Thai flue-cured tobacco to Japan. In the years following the joint venture made substantial investments in a re-drying factory and in a network of buying stations in the north of the country. The big leap forward came in 1974 when Philip Morris Inc. became interested in Thailand's oriental tobacco. On the basis of a sample purchase it undertook to buy all the Thai Turkish tobacco that Thailand could produce, provided increases in quantity and price were moderated to avoid violent fluctuations. It urged Adams International to:

a) invest in proper manipulation machinery to extract the sand from the tobacco;
b) maintain and improve quality and grade standards;
c) ensure availability of land and farmers so that production could be increased yearly;
d) supply Philip Morris on a first priority basis; and
e) assure that there would be no big fluctuation in price.

The company responded by installing two new lines of manipulation machines. It also introduced tobacco to several self help settlements in the North Eastern provinces using new varieties from Greece and Turkey.

The production programme foresaw that each region would be supervised by a fully qualified agronomist under the over all direction of an agriculture manager. Under the agronomists were head village inspectors who supervised the work of a team of village inspectors. Each village inspector was provided a motorcycle. He was trained to advise farmers on planting, curing, baling and grading. He had also to motivate farmers to keep up their work so that prior efforts and investments in agricultural materials would not be wasted for lack of follow up. Thus the village inspector was much more than an adviser.

All the materials needed for a tobacco crop were supplied to farmers on an interest free credit basis; the cost was deducted from the sales proceeds. This was necessary to ensure that the correct fertilizers and insecticides were used. The inputs required could be as many as 16 ranging from fertilizer, lime, insecticides and fungicide to spray pumps, water cans, plastic sheets, twine, needles and burlap.

Tobacco was delivered by the farmers in a dozen grades, each with a different price. New farmers were taught to grade carefully and pack each grade in bales of 12 to 15 kgs. With 40,000 farmers receiving inputs and delivering product, record keeping was a major accounting task.

Crop diversification was strongly recommended. Optimum tobacco production called for a rotation that fed back into the soil the nutrients tobacco used up. Growing other crops also supplemented the farmer's income and reduced the risks of dependency on a single product market.

The line of authority in the company went from the board of directors through its executive chairman to factory line managers, and two field managers. Reporting to the field manager was a regional manager responsible for a network of buying stations. Reporting to him was a station manager who supervised the warehousing, cleaning and packing facilities. Under the station manager were head inspectors each responsible for farmers in five to seven villages. The vital last link in the managerial chain was the village inspector who dealt with the farmers on a daily basis. In 1983 the company had 600 village inspectors recruited from successful farmers. Instructions to farmers were prepared in pictorial form.

The payoff has been substantial for everyone involved. The buyers, primarily Philip Morris, obtained access to a supply of tobacco for blending purposes at a price 30 percent less than that of their traditional sources in Turkey and Greece. Adams International, the Thai joint venture began

Table 2.10 Operating results;
Adams International 1975, 1980, 1983

	1975	1980	1983
Tobacco purchased (tons)	760	3,223	4,589
Tobacco exported (tons)	580	2,660	4,017
Yield (percent)	76	83	87
Price received ($ per kg.)	1.45	2.21	2.31
Cost ($ per kg.)	1.28	1.98	2.20
Margin (percent)	11.7	10.4	4.8
	(. . . $ thousands . . .)		
Total sales	840	5,878	9,280
Profit	98	612	442

operations with four employees; these grew to 700. Oriental tobacco amounted to about half of its sales and contributed more than half of its profits. Table 2.10 shows that the yield of export quality tobacco increased steadily, from 76 percent in 1975 to 87 percent in 1983. Sales also rose, but profits were less in 1983. In that year Greece and Turkey received more favourable treatment from the EEC and Thai tobacco lost much of its price advantage on that market.

The company put its total investment for the years 1974-83 at about $3 million. This included machinery, warehouses, and buying stations. Of this investment, about 20 percent went on extension. Operating the buying stations and the salaries of village inspectors cost about $500,000 a year.

From the farmer oriental tobacco required more and harder work than many crops but it brought higher returns. While the income from growing rice was about $260 per ha. gross, tobacco produced a net of $1,140. It also provided off farm employment for children stringing the leaves.

BUNYORO COOPERATIVE GROWERS' UNION, UGANDA

Cooperative ginneries played an important part in the expansion of small holder cotton production in Uganda in the 1950s and 1960s. It became a major source of cash income for the rural population and of foreign exchange earnings for the country. Bunyoro Cooperative Growers' Union was one of the first five unions of village level societies. It was registered in January 1954 with 23 member societies and share capital of $2,000. That year, it only supplied to members the hessian sheets used for sorting and bagging their raw cotton.

Towards the end of 1956 government purchased a private ginnery at Masindi and leased it to the Union for two seasons. The surpluses made in those years were capitalized. The Union was able to put down one third of the price of the ginnery in 1958 and so became its owner. It spent $7,000 on improvements. After the 1958/59 season, it was able to distribute a small part of its surplus in cash; the greater part, as with earlier surpluses, was capitalized as "Bonus Shares". In 1963, a second ginnery at Hoima was purchased directly from the owners on a five year credit basis without a government loan. However, the cotton cooperatives were backed by a government policy of concentrating small holder sales through them. From 1963 the Bunyoro Union was assigned 78 percent of the cotton grown in the area it served and from 1967/68 100 percent. Over this period it undertook a cautious programme of improving and expanding its ginning capacity. It also helped members market other crops such as coffee, tobacco and groundnuts. From 1963 all its staff were Ugandans.

The location of the Union's ginneries and the technology used were determined by decisions of the previous private owners. Initially they were under used. Later cotton production increased to the point that the Union's plants were barely able to cope. Two shifts were operated and the season was extended from the normal 120 to 130 days to 170 days. In 1964 the Platt single roller double action gins at Masindi were replaced by new Middleton double roller gins. Old machinery was gradually replaced at Hoima also.

Under the cotton cooperative system in Uganda Union contacts with growers were through the village societies of which they were members. The Union sold fertilizers, insecticides, cotton sprays and sprayers, but had no staff to guide farmers in production techniques. For this the government extension service was responsible.

The cotton lint and seed separated by ginning were sold directly to the Lint Marketing Board at predetermined prices. So the Union had no sales risks but it had to await the orders of the Lint Marketing Board before dispatching consignments. This could create a storage problem. It also had to wait a long time for payment. These delays added significantly to its costs.

The Union operated on a margin set by the price payable to growers for their cotton through the village society and that it received from the Marketing Board for lint and seed after ginning. The Bunyoro Union had lower costs than other cotton processing cooperatives in Uganda and its management was content to keep within the margin allowed. It was the cooperative department of the government that pressed for the institution of a standard budgetary control system from 1968. The Union paid higher wages to its employees than those set by the government wage board and those otherwise prevailing in its area, but a productivity incentive scheme suggested by the cooperative department was not implemented. Management was considered a routine operation with little use made of leadership training opportunities. Perhaps it had grounds to be complacent. In 12 years it had raised its meagre share capital to $200,000. This came from capitalization of annual surpluses. Direct subscription of new share capital was hardly $2,500. Quality wise, moreover, the bales ginned and pressed by the Union were above average for the country.

UNILEVER IS, TURKEY

Unilever is one of the largest agricultural processing transnationals with a strong position in oil seeds. In Turkey, as in many other countries, it has pioneered the processing of oil seeds into margarine, other cooking fats, and soap. In 1983 its sales there of Sana margarine amounted to 108,000 tons and were the largest in the world for one brand.

Effective Unilever interest in Turkey began with an exploratory mission in 1949. It foresaw an important market for 'a cheap and wholesome substitute of butter and clarified butter as both these articles are outside the reach of the large majority of the population.' The raw material would be domestic cotton and sunflower seed, supplemented by imports of soybean oil if needed. There was no significant competition in sight. Unilever proposed to the government the establishment of a plant, stressing the following advantages to the economy:

a) creation of a new outlet for domestically grown oil seeds;
b) making available to consumers products of high nutritive value;
c) low selling prices would help reduce the rapidly rising cost of living;
d) high value olive oil could be freed for export.

The government welcomed the proposal and encouraged Is Bank to become a partner with a 20 percent share provided in local currency. The value of a transnational was demonstrated by the speedy deployment of technical, raw material supply, marketing and promotion experts. Later they were replaced by local personnel. By the 1980s Unilever Is was essentially Turkish.

Sales expanded rapidly. The population rose from 21 to 30 million from 1950 to 1965. Unilever was able to keep the price of its margarine and vegetable ghee well below that of butter and natural ghee because the cost of its raw materials did not rise as fast. Thus in January 1962 fresh salted butter Trabzon type cost $1.44 per kg. as compared with Sana margarine at $0.67. Olive oil cost $0.89 per kg. while Unilever Is sold Vita at $0.62 to $0.64 per kg. These price advantages were very important to Turkey where the average income was then only $180 per annum. Low costs also helped Unilever Is during periods of government price controls. Maximum prices based on the production costs of its Turkish competitors still allowed Unilever Is to make a profit.

Though it faced problems over supplies, taxation and legal definition of its products, Unilever did very well in Turkey. Sample financial results are summarized in Table 2.11. Between 1951 and 1965 issued share capital rose from $1.79 to $4.62 million of which 80 percent belonged to Unilever, the rest to the Is Bank. In the same period Unilever earned $5.45 million in dividends and the Is Bank received $1.36 million. Transfers out of the country amounted to about a quarter of total profits. The Turkish Government took more than this in taxes.

One of the factors to which Unilever Is attributes its success is efficient purchasing of its raw materials. Cotton seed oil was a by-product of the cotton industry in southern and western Turkey. Unilever Is bought the oil from the crushers. Sunflowers were grown specifically for the oil seed market, primarily in Thrace. Again, Unilever Is bought mainly from local crushers, also seed from a growers' cooperative which was then crushed at a mill on contract. Most of the time the government maintained a guaranteed minimum price for sunflower seed paid to farmers on delivery to the cooperative.

Table 2.11 Summary trading figures: Unilever Is, 1956, 1965, 1983, 1984

	1956	1965	1983	1984
	(. . . $ millions . . .)			
Net sales	27.22	41.61	100.28	126.50
Gross margin	4.29	6.78	24.93	29.61
Costs				
Advertising	.35	.98	1.04	.88
Marketing	.63	1.17	.41	1.29
Factory and general	.70	1.62	6.24	8.35
Head office, research, pensions	.20	.61	2.00	2.71
Total costs	1.98	4.38	9.69	13.23
Profit before tax	2.31	2.40	15.24	16.28
Profit after tax	1.78	1.54	9.09	8.99
Gross capital employed	9.07	16.59	31.42	40.04
	(. . . . percent . . . .)			
Costs in relation to sales				
Advertising	1.3	2.3	1.0	0.7
Marketing	2.3	2.3	0.4	1.0
Factory and general	3.0	4.0	6.2	6.6

However, farmers could commit their crop by contract to a crushing mill at a discount from the government price in return for a cash advance. Similarly, Unilever Is advanced funds to two cotton seed and two sunflower seed crushers on the understanding that it would receive all their oil when it was ready. Thus Unilever was not so much a price setter for farmers as a consistent sure outlet for the semi-processed product. Output of sunflower seed rose from 50,000 to 700,000 tons. Unilever Is also had a continuing influence on quality. It applied a consistent set of deductions if oil failed to meet established standards of colour and freedom from fatty acid.

Concerned always about the supply of its raw materials Unilever promoted a new company in the early 1980s together with the Thrace oil seed growers' cooperative and the Interstate Seed Co. of America to produce hybrid sunflower seed in Turkey.

Because of the importance of quality control at the consumer level, Unilever Is maintained its own distribution network; 75 percent of sales are directly to retailers. The frame for this is five wholesale depots and 11 sub depots. Salesmen go out from these to take orders. Vans follow them the day after with deliveries. Stocks unsold after two months are taken back and remelted. Experience also shows that this system of distribution can be cheaper. The conventional mark-up has been five percent for the wholesaler, 10 percent for the retailer. Costs at the direct distribution depots have often stayed within the two to three percent range.

EXPORTS OF BEEF BY AIR FROM CHAD

For large parts of the Sahel countries of Africa livestock is the main agricultural resource. Large herds are moved along traditional routes to take advantage of seasonal grazing and meet watering requirements. The longstanding outlet for such animals as were marketed was trekking on foot to the humid coastal areas where prevelance of the tse-tse fly inhibited local production. Marketing in this way involved substantial costs and quality losses. More important, animals had to be kept until they were four or five years old if they were to withstand the journey.

In the 1950s and 1960s the possibilities of using

air transport to carry meat from distant production areas to higher priced market areas became recognized. Governments of the Sahel countries mobilized aid resources to construct abattoirs that could be the base for such transport. The Farcha abattoir near N'jdamena in Chad is a notable example of this. It was conveniently situated near the international airport. A small number of locally based companies sent out agents to buy cattle from range herdsmen. Holding grounds were fenced and watered. Here animals rested and were conveniently available for slaughter when transport of the meat had been arranged. Carcass quarters were then exported in the cargo holds of scheduled passenger airlines flying to African cities such as Douala, Libreville and Kinshasa. This seemed an effective solution to an intrinsically African problem; how to move meat from the savannah where it was plentiful and cheap to the coastal cities where it was scarce and expensive, and so benefit the peoples of both regions.

The abattoir was designed and equipped to meet the standards required by the markets served. Cold rooms were provided for chilling and holding carcasses while awaiting transport. In 1970 the volume of beef exports reached a peak of 14,000 tons.

Though incurring high transport costs the Chad exporters developed their markets through regular deliveries and consistent quality. This was feasible so long as their requirements were only a small part of the total supply. However, the productive capacity of the Chad grazing area could not easily be expanded in response to market demand. It was especially vulnerable to drought. This showed up eventually in Chad. Procurement prices rose sharply. The exporters were left with too narrow a margin to cover air transport of the meat.

In 1974-75 the quantity of beef exported had fallen to around 7,000 tons. It still brought in $10 million to a country with only meagre resources. These returns were reflected back to the original producers as higher prices for their cattle than if they were trekked out on foot. The major markets then were

Congo-Brazzaville, Central African Republic, Zaire, Ghana, Libya and Kuwait. Supply systems in Chad had deteriorated, however, because of discontinuities due to the drought and other factors. The holding grounds had been abandoned. The six or seven exporting firms sent out agents to buy cattle on local markets and through intermediaries who had contacts with herdsmen when they obtained a profitable order; but continuing contacts between exporters and producers were limited and organisation to hold cattle conveniently ready for purchase when export opportunities arose was lacking. Prices tended to rise sharply when an exporter tried to buy and fell again thereafter.

It appears therefore that provision of processing facilities on a custom basis by an abattoir in municipal ownership and export marketing by competing local firms is not a sufficient structure for optimizing productivity. One or more of the exporting enterprises must have access to the resources needed to establish holding grounds and maintain continuing relations with a group of producers.

BOTSWANA MEAT COMMISSION

This enterprise began thirty years ago as a private slaughterhouse, known as the Lobatse Abattoir. In 1958 it was acquired by the Commonwealth Development Corporation (CDC). The Meat Commission was established in 1964. Its purpose was: " ... to secure that so far as is reasonably possible all livestock offered or available for sale in Botswana to the Commission are purchased and that the prices paid therefore are reasonable". Agreement was reached with CDC to purchase its 50 percent holding and convert it to a loan. Thus the abattoir became wholly government owned.

In spite of interruptions in access to external markets and internal shocks such as drought and foot and mouth disease, the BMC has given Botswana cattle producers confidence in a continuing market for their animals at a reasonable price. In the past 15 years its throughput has more than doubled and prices paid have risen at a rate of 10 percent per year. The BMC has also been a major contributor of tax revenues to the government.

Livestock numbers in Botswana are uncertain, likewise the rate of off take. Nonetheless, it is high compared to many African countries. This is attributable to the high prices which BMC has paid, a relatively efficient system for getting livestock to the abattoir, an increasingly commercial attitude on the part of the tribal producers, and the small but important freehold farming sector.

Many of the freehold farmers grow on or fatten animals bought from tribal farmers as well as raising their own. Government finishing ranches have been established to provide a parallel service, but this is on a relatively small scale.

In order to maintain a steady day to day throughput at the abattoir BMC operated a quota system. Each supplier who wished to send cattle for slaughter had to apply for a quota and was penalized for non-fulfilment. Agents established themselves to help arrange quotas, organise rail transport and look after the producer/clients' interests at the abattoir. They operated on a commission basis at competitive rates. A cooperative union acted as agent for cooperative societies who found livestock marketing an excellent basis on which to establish themselves.

BMC purchased livestock by dead weight and grade. To enable producers to plan ahead it issued a yearly schedule setting out prices for each grade for each four weekly period. At the end of the year any surplus was distributed in the form of a bonus to suppliers.

The policy of the Commission has been to do as much processing as possible within Botswana. Edible and inedible offal, bloodmeal, bonemeal, carcassmeal, tallow, horns, hooves, ear and tail hair, ox gall were sent to specific markets. Hides once exported in a raw, salted form were exported as wet blues. Pet food canning was added in 1981.

Sales in continental Europe and South Africa have

been effected by local agents. In the UK a joint company, Allied Meat Importers (AMI) was set up in partnership with a British firm of meat traders. The whole operation, AMI and ECCO Cold Store Ltd., was managed by BMC itself through Botswana Meat Commission (UK) Holdings. Supplies from other sources were also handled so that, if supplies from Botswana were interrupted, BMC (UK) continued to operate. Its profits were remitted back to Botswana for distribution to producers. Table 2.12 summarizes BMC income, expenditure and appropriations in 1969 and 1982.

Foot and mouth disease is one of the major constraints to the development of meat exports from Africa. European importing countries are acutely aware of the risks of importing exotic strains to which their cattle would be highly susceptible. That Botswana has been able to overcome these fears is largely due to the efforts of its Director of Veterinary Services. The country was divided into zones separated by cordon fences. A permit was required to cross from one control zone to another and this could only be done at certain points where a quarantine was maintained. The continued willingness of importing countries to accept imports from Botswana was proof of their confidence in the ability of the Botswana Veterinary Department to monitor the disease and act promptly should an outbreak occur.

The Meat Commission was one of Botswana's major employers. Until the development of large scale mining BMC was also the main source of income for the country and of export earnings.

Operating in a complicated and extremely competitive market BMC has consistently provided an attractive outlet for Botswana's livestock and hence encouraged the development of the industry. Livestock raisers' incomes have benefited correspondingly. Commercial viability has been preserved by the hiring of professional staff and by a favourable relationship with the Botswana Government. It has been left free to operate in a commercial fashion. At the same time it has been helped greatly by the government through the maintenance of livestock disease controls and

Table 2.12 Income, expenditure and appropriation accounts: Botswana Meat Commission, 1969 and 1982

	1969	1982
	(. . $ thousands . .)	
Income		
Sales of meat, meat products, services	14,343	102,190
Interest receivable	35	52
Increase in stocks	176	5,444
Miscellaneous	10	665
Brought forward	17	198
Total income	14,581	108,549
Expenditure		
Payments to producers - initial	9,302	52,971
- bonus	1,095	7,833
Purchases of meat in London	-	767
Freight, storage, levies, sales	1,276	17,366
Processing and administration	1,343	15,637
Depreciation of fixed assets	382	1,356
Taxes	826	9,275
Total expenditure	14,224	105,205
Net income	357	3,344
Appropriations		
Transfer to capital loan redemption reserve	143	249
Appropriation to capital reserves	214	396
Transfer to stabilisation fund	-	1,849
Carried forward	-	850

lobbying in Europe for favourable terms of entry to its relatively high priced markets.

A possible criticism is that BMC did not organise purchasing nearer to the tribal producers. It bought from, and paid eventual bonuses to the man who delivered the animals to its abattoir. Producers' cooperatives were organised to undertake delivery. Nevertheless, the bulk of the tribal livestock raisers received prices much lower than those paid by the Commission. On the other hand, the farmers and wholesalers who bought tribal cattle for fattening before supplying them to the Commission were a major factor in its ability to export quality meat. It can also be that trekking livestock over long distances through arid country is better done by people who own them than by employees of a distant parastatal.

SWAZILAND MEAT CORPORATION

Swaziland Meat Corporation (SMC) originated from a 1964 agreement between the Government of Swaziland and the Imperial Cold Storage Co. of Pretoria. The original objectives were to process meat for South African and local markets and reduce overstocking on traditional grazing lands. In addition to an abattoir, the Corporation operated a canning plant to take up supplies when foot-and-mouth disease prevented fresh exports, a chain of retail shops and holding grounds.

The authorised share capital of $2.8 million was provided by the South African partners. Fifty thousand shares were allotted to the government in trust for the Swazi nation. Later this share was expanded to half by means of a loan from the company later repaid out of dividends. The abattoir was granted a monopoly of exports, extended in 1980 for a further ten years.

Currently the commercial partner in the Corporation is SMC Investments, a subsidiary of the Anglo-American mining company. It provides management which is charged against operating costs, not for a fee. Policy making is vested in a board of eight directors, four representing SMC Investments and four the Swazi nation, nominated by the King in Council. The latter take little part in policy formulation. The chairman is nominated every two years by the commercial shareholders, subject to approval by the ministry of agriculture and cooperatives.

Increased demand for slaughter stock followed the establishment of SMC as an export abattoir. The average number of local stock slaughtered between 1977 and 1982 was 35 percent higher than the quantity exported live prior to SMC's establishment. SMC has therefore made an important contribution in raising off take and reducing overall stocking intensity on Swazi national land.

In 1977 SMC introduced a system of minimum guaranteed prices on a live weight and cold dressed weight basis, with prices quoted for three grades, fat, medium, and compound. To help ensure a regular supply of stock SMC set up a network of buyers who would contact livestock owners when they took their stock to dip tanks. This replaced to a considerable extent purchase at open auction.

At one time SMC had several farms for raising cattle and holding them prior to slaughter. In 1983 these activities were confined to a feed lot utilizing by-products from a citrus factory. Several holding and fattening ranches were operated by the government with credit arrangements designed to attract cattle from over-stocked national lands. The main buyers from these ranches were civil servants and other investors.

Under the Lomé Convention, Swaziland had a quota of 2,000 tons on the EEC market. The plant was brought up to EEC standards in 1975. However, these standards kept rising and being interpreted more strictly. Insufficient funds have been available for alterations to keep pace with the changing requirements. In January 1982 the plant closed for four months to enable essential work to be carried out before the visit of the EEC inspectors. The cannery was closed for export at around the same time, and could only produce for the domestic market. In spite of these deficiencies SMC's products have a good reputation for quality and reliability in the UK and West German markets, and it is most important that Swaziland retains this export market.

As of 1983 prices at auctions tended to exceed SMC's floor price and its intake had turned down. This may have been related to the delays in re-equipping its facilities. Insofar, of course, as it was due to competition from buyers for other markets there was only advantage to producers.

Under consideration in Swaziland is the conversion of the Corporation into a commission on the lines of that of Botswana as a basis for raising the $10 to $12 million needed for a new plant. Whether it would be able to attract the volume of supplies needed to justify such an investment is an important issue.

LA FAVORITA, ECUADOR

Santa Domingo de los Colorados is one of the growing market centres of Ecuador. It is situated at a point where roads from the north and south of the productive coastal plain meet to pass over the Andes to Quito. The population of Santo Domingo is growing at a rate of 10 percent per year. The volume of produce passing through is rising even faster.

In the early 1970s, responding to pressure from livestock raisers, the government established a mixed government/producer enterprise to operate an abattoir to serve the Santo Domingo region. Though a number of livestock raisers had a direct financial interest the plant was unable to compete with the traditional livestock and meat marketing channel which it sought to circumvent.

The traditional system was founded on livestock wholesalers known as "introductores". They bought animals directly from farmers by private treaty, or at periodic livestock markets. They used one or another municipal slaughterhouse to have them slaughtered. Thence they delivered the meat to urban retailers and other parts of the animal to specialized buyers. Running through this system was a patronage/credit thread that made it difficult for an outsider to break in. The wholesaler dominated the staff of the municipal abattoir he used, threatening to move his business elsewhere if it raised its charges. His sales were assured through credit ties to retailers. Held together by direct personal contacts this system operated at very low cost.

Retail price controls had, for a number of years, inhibited the payment in Ecuador of incentive prices for quality. However, supermarket chains catering to middle and higher income consumers such as La Favorita had tacitly been allowed to offer quality meat at higher prices. To be assured of continuing supplies for this middle and upper income consumer market, La Favorita set up its own abattoir near Santo Domingo. Its objective was to make contracts directly with livestock producers who would raise animals meeting the quality standards it had in mind. In this it never succeeded. It had to compromise on engaging four of the established "introductores" to assure the throughput of the abattoir. Livestock raisers in its supply area were, nevertheless, producing animals of the quality it required in marked contrast to the prevailing practice, and received higher prices in consequence.

Thus the retail enterprise, by establishing its own processing facilities had been instrumental in promoting production of higher quality livestock in its supply area. While still unofficial the Santo Domingo operation was regarded as a model for the future development of the livestock economy of Ecuador.

JAMHURI TANNERY, TANZANIA

This tannery was registered as a company with four shareholders in 1964. It operated on its own plot at Mkuu in Rombo district. It processed hides into sole leather, goat and sheep skins into lining leather. It had also tanned game skins. In 1972 the shareholders had increased to 10. Paid up capital was $6,225. Assets in buildings, machinery, stock in trade and cash amounted to $35,412. The enterprise was run by a general manager who had been with the company from its outset. There was a sales department with its own manager. He supervised outside sales and also ran a shop in Moshi with one attendant.

In 1972 the plant processed a monthly average of 1,300 hides into sole leather and 4,000 skins for linings. It operated two revolving drums to speed up tanning. This was vegetable based requiring about one ton of wattle extract per month. Its operating costs are set out in the income and expenditure account. (See Table 2.13)

Hides and skins were purchased from butchers and brought in by buyers from the rural areas. Thus, contacts with livestock producers were primarily indirect. However, the value of the hide or skin was a significant element in the price at which producers could sell their animals, particularly those coming from dry range or kept for a number of years to give milk.

Tanned hides and skins were sold to shoemakers and repairers locally and in Kenya. Overall, however, Tanzania was a net importer of leather. Its livestock population was considerable, but much of its hides and skins output, which was estimated at 1.36 million pieces in 1972, went out of the country in sundried or wet salted form. Demand for leather shoes in Tanzania was expected to increase steadily from rising income and population and greater awareness of the need for footwear. Supplies of raw hides and skins were ample and expected to increase. Jamhuri Tannery was considered to be well managed. With this background it decided to undertake a major expansion. There were two main competitors, Tanzania Tanneries and Himo Tanneries. These firms used chemical tanning processes and were more mechanized. Jamhuri proposed to follow their lead.

It was proposed to expand throughput at Jamhuri to 480,000 kg. annually of hides and 120,000 skins to produce 529,000 kg. of leather. 3,000 hides and 6,000 skins would be chrome tanned monthly for export. The

Table 2.13 Income and expenditure account: Jamhuri Tannery, 1972

Income	$
Sale of 124,000 kg. hides at $1.03 per kg.	128,544
Sale of 38,400 kg. skins at $2.17 per kg.	83,328
Total gross income	211,872
Less - 15,600 hides at $5.37 each	83,792
48,000 skins at $1.08 each	51,840
Gross operating margin	76,260
Expenditure	
Salaries for 17 workers	17,883
Processing materials	14,341
Diesel oil, wood for fuel	12,805
Transport charges	2,195
Maintenance of buildings, machinery, etc.	2,751
Rent	2,049
Office expenses, telephone, etc.	176
Workers' insurance	1,788
Total operating cost	53,988
Net operating income (before interest, depreciation and taxes)	22,272

tanning process used currently and that envisaged were as follows:

1. Raw hides and skins were soaked in water for 72 hours, then put into a pit containing sodium sulphide for a further 96 hours. This treatment could be shortened to 24 hours if revolving drums were used.
2. The next step was to remove the hair. By hand an experienced shaver could deal with 50 hides a day. With a shaving machine, one person could shave 100 hides per hour.
3. The hides were then split into half and

soaked in a pit containing soda solution for 72 hours. A revolving drum would reduce the time to 12 hours.

4. The hides were washed in clean water to remove all soda solution; then all meat was removed.
5. The hides were then soaked in drums containing sulphuric acid solution for 45 minutes, then removed and washed in clean water.
6. The hides were then soaked in a pit containing wattle extract solution for 72 hours.
7. The hides were then put in revolving drums containing preservative chemicals for 48 hours.
8. Hides were then washed in clear water to remove chemicals and put in revolving drums containing fish oil and china clay for 45 minutes to make the leather supple. Skins were dried for ten days, after which they were ironed and became ready for use.

With the chrome process hides and skins are only half tanned. They are mainly used for shoe uppers. The process follows steps 1 to 4 in tanning. Then instead of soaking in sulphuric acid solution, hides and skins are soaked in chrametan for 24 hours. Then they are washed in clean water and suspended on lines for drying in shade for three days. Finally they go into a buffing machine and then a glazing machine. They are then ready for export.

The expansion could be accommodated on the existing site. It would provide for 26 jobs, an increase of nine on the existing staff. Capital expenditure on machinery and other items would be $32,630. Additional shareholders would be brought in contributing $8,240 of equity. A three year loan of $24,400 would be sought from the Tanzania Rural Development Bank (interest rate 7 1/2 percent annually). An overdraft to finance the purchase of much larger quantities of raw hides and skins would be sought from the National Bank of Commerce.

On the basis of the projected capital outlays, the operating costs anticipated, and the income expected, the project would bring in a net income after tax of $14,300 in the first year, rising to $16,900 in the tenth year. This assumed operation at 98 percent of capacity.

The following estimates were made as a basis for financing the proposed expansion.

	$
New investment	
Four roving drums	11,700
One lister diesel engine 25 hp.	2,200
One buffing machine	7,160
One glazing machine	2,770
Building expansion and fitting	8,800
Total	32,630

Depreciation would be allowed for as follows:

	Estimated value	Depreciation percentage	Depreciation per year
Existing assets	$		$
Buildings	8,200	5	410
Plant & machinery	2,400	12 1/2	300
Furniture & fittings	2,440	20	488
Vehicles	7,320	25	1,830
New investment			
Roving drums	11,700	10	1,170
Lister engine	2,200	12 1/2	275
Buffing machine	7,160	12 1/2	895
Glazing machine	2,770	12 1/2	334
Total depreciation			5,702

The depreciation rates used were those acceptable to the Government of Tanzania for tax allowances.

Table 2.14 Projected annual income and expenditure after expansion: Jamhuri Tannery

	$
Income	
Sales	
sole leather 188,160 kg. at $1.04	195,043
chrome leather 705,600 ft. at $0.30	211,680
lining skins 23,520 kg. at $2.08	48,892
chrome skins 35,280 kg. at $1.77	62,385
Total gross income	517,900
Less	
60,000 hides at $5.30 each	318,000
112,000 skins at $0.87 per piece	97,440
Gross operating margin	102,460
Expenditure	
Salaries - general manager $244, sales manager $183, shop manager $85, 6 staff at $48.80 each monthly	9,656
Wages - 16 factory workers at $48.80 each monthly	9,370
Provident fund contributions	1,408
Processing materials (assuming capacity operation)	23,900
Fuel - 90 litres per day at 17.6 cents per litre for 26 days per month	4,942
Transport - $36.60 per day for 15 days per month	6,588
Maintenance and repairs - estimated at 10.47% of cost of equipment	5,733
Rent	1,950
Office and other expenses	2,970
Workers' insurance	2,043
Total operating cost	68,560
Operating income	33,900

Income and expenditure figures projected for the tannery after the expansion are presented in Table 2.14. Prices were based on those current in local and international markets. On this basis the tannery expected to make an annual net income of $14,300.

Its loan repayment schedule was as follows:

Year	Principal outstanding	Interest payable	Loan instalment	Total payable
	$	$	$	$
1	24,375	1,830	8,125	9,935
2	16,250	1,220	8,125	9,345
3	8,125	610	8,125	8,735
		3,660	24,375	28,035

Profit and loss and cash flow forecasts showing that the loan interest charges and repayment instalments could be handled without difficulty are shown in Chapter 5. There also calculation of the financial rate of return on the investment is demonstrated - over 50 percent. These projections assume of course, continuation of the same purchasing and sale price relationships, volume of business, and costs. These are subject to many influences. The degree of risk of an adverse shift in income/cost relationships should be reflected in the rate of return projected.

THE BROILER BOOM IN LEBANON

Considered spectacular in the 1960s was the growth of the broiler industry in Lebanon. This was a private enterprise initiative with a minimum of institutional support. In a few years poultry production in Lebanon switched from a side activity of general farmers to a specialized industry of some 100 broiler raisers. The method of sale to the consumer also changed from live birds to eviscerated birds packed in ice, and later to frozen carcasses ready to cook. Processing had a strategic role in this.

The starting point was the arrival in Beirut of imported broilers demonstrating the efficiency of American and European enterprise. Technical information was available through published material and individual contacts. Many businessmen saw the possibilities of applying this technology in Lebanon and of exporting broilers to the oil rich countries of the Middle East where they already had commercial contacts.

The government helped by providing protection against competing imports. It established the Fanar Institute to work on poultry diseases. Vaccines became available free of charge. This reduced the risk of losses due to epidemic disease formerly a deterrent to concentrated poultry production.

Over the years 1954-69, poultry meat output in Lebanon increased ten times. Poultry rose to 20 percent of agricultural output by value and a similar proportion of exports. This phenomenal growth began on imported chicks and feed stimulating in turn the establishment of domestic hatcheries and feed mills.

Increased specialization was promoted by periodic profit squeezes. General poultry farms were replaced by layer, broiler and hatchery enterprises backed by specialized feed and veterinary services so permitting production at lower cost. While the average cost of producing and dressing a broiler was about 60 cents, prices ranged from 55 to 95 cents. A downward trend in prices towards 1970 reflected the impact of competition and steadily increasing efficiency in production and marketing.

At this stage most broilers were processed on the farm. The cost around 1960-61 was estimated at 4.5 cents per bird. The first large scale poultry abattoir with modern equipment was established at Zahle around this time. It could handle 8,000 broilers in 10 hours. Under pressure of its competition the cost of dressing broilers had fallen to an average of 3.3 to 4.4 cents per bird in 1968. Use of refrigeration also became general during this period.

Marketing channels became more specialized also. Originally broiler raisers brought their own output to market. They dealt directly with retailers and often had to wait a few days until accounts were settled. The difficulties of this system of marketing gave rise to a class of broiler wholesalers who marketed growers' dressed broilers for a margin of 3.3 to 5 cents per kg.

Integrated operations based on credit from feed companies and hatcheries developed during this period. The ease with which producers could obtain chicks and feed on credit, amounting to around 90 percent of the cost of production, was a main factor in promoting periodic phases of over supply and consequent low prices. Some producers with access to freezer storage could hold stocks off the market if prices were considered too low. Those who had entered the business to take advantage of high prices and lacked a secure outlet came under pressure to meet their credit obligations. They had to lower their prices to clear their output. This accentuated a downward trend in prices.

While the larger wholesalers had a strategic role in developing the export market, it does not appear that they were instrumental in transmitting technology to the producers with whom they dealt. Nor was the government; it helped with disease control, provided protection against competing imports and in 1967 issued decrees designed to control export quality. The initiative in this development appears to have come mainly from persons outside agriculture entering broiler production with up to date technical information and commercial capital.

JAMAICA BROILERS

Jamaica Broilers, the largest and oldest broiler processor in Jamaica, was founded in 1958 by three men who were importing iced broilers. When their imports reached 100 cases of 27 kg. weekly, they decided to produce the chickens locally.

From the outset, they adopted a contract grower system. Agreements were made with farmers who would build a broiler house to company specifications, purchase the equipment and look after the chickens for the eight to nine weeks required to reach marketable size. The company supplied day old chicks, feed, medication and technical support at low cost. Available were the services of two veterinarians, a poultry nutritionist and an eight person field advisory team. The chickens remained the property of the company. The growers were paid under three heads: rental, performance and price. The rental payment was based on the cost of building and equipping the broiler house. There were five levels of rent depending on when the house was built and the materials used. Under this head the contracted grower received a weekly payment whether or not there were birds in his house. The performance payment was based on the average live weight and feed conversion rate recorded for the flock. It rewarded efficiency in managing the day old chicks and the feed supplied. The third category of payment responded to the pricing policies of the government. For many years there were strict price controls. When an increase in the retail price of frozen broilers was authorised a percentage of this increase went to the contract grower. This payment could be as much as 90 cents per bird.

At first most of the inputs were imported - technical know how, day old chicks, balanced feed, medicines, equipment and finance. Then a chick hatchery was established in Jamaica, then a feed mill. Under the pressure of foreign exchange restrictions the company was obliged to use more and more local ingredients. It flourished as demand increased with the change over in retail food marketing from the local grocery store to supermarket style operations. Sales promotion has been very good. While few people in Jamaica would recognize the name Jamaica Broilers, everyone knew "The Best Dressed Chicken" its successful brand name.

Led by Jamaica Broilers, broiler production in

Jamaica grew at a rate of 15 percent annually from 1965. In 1980 the firm marketed 16,700 tons out of a national production of 30,000. At this time Jamaica Broilers had 260 growers with an average of 14,000 birds on their farm.

A feature of Jamaica Broilers was that the various participants in its operation - farmers growing chickens under contract, contractors providing transport services and direct employees in the processing plant, had shares in the company. In the mid seventies one of the original owners wished to capitalize on his investment. The American partner in the feed mill - Central Soya - withdrew. Shares were offered to all employees and contractors. Over 90 percent of the people eligible took up the offer. Shares were paid for over five years by deductions from earnings. If employees or contractors left the company, their shares were to be sold back to an employees' trust. In a country with strong labour unions employee participation in company ownership has been an important factor in its continuing success. It has also been an incentive to growers in addition to that available under their contract.

By guaranteeing a market outlet and rewarding efficiency the Jamaica Broilers contract was a potent force for optimizing take up of the technology made available.

CHAROEN POKPHAN GROUP, THAILAND

Backyard production of indigenous chicken is still the main source of chickens for consumption in rural areas of Thailand. However, production, processing and marketing on a commercial scale has shifted to large vertically integrated companies with sophisticated knowledge of disease control, feeding, processing and marketing.

The role of the C.P. Group has been to make advanced technology available to local farmers in a form that they can apply with a minimum of risk. The tendency has been for more and more formerly independent producers to negotiate with firms such as the Charoen Pokphan Group to raise chickens in return for a fixed fee or wage contract. C.P. was the first firm to initiate this type of contract in Thailand. It provided producers with an agreed number of chicks and the feed, drugs and veterinarian services required at no cost. Producers who had no chicken houses could build one or two houses to hold 5,000 to 10,000 birds each by obtaining loans from commercial banks negotiated for them by C.P. In return, they signed a five year contract to raise chicken under the close supervision of its officers. Producers were paid between five to seven cents per marketable broiler. A penalty was imposed if the broiler's weight was lower than that specified in the contract; a premium was paid for extra weight.

Supervision of wage contracts could be quite costly in the absence of proper organisation. For this reason it was not in general use.

There appeared to be no difference in the productive efficiency of price guaranteed growers or of piece rate growers. The shift towards flat fee contracts was attributed to risk aversion and transaction costs. Under a piece rate contract the integrator bore most of the risk.

Vertical integration was necessary for a young industry such as broiler raising because it was not possible to purchase inputs or services from other suppliers; they were not yet available. Since vertical integration usually required a very large initial capital requirement, an optimum scale of production was needed at each stage of growth so that economies of scale could be achieved and fixed costs could be covered. This was the basis of the success of the C.P. Group which controlled 44 percent of the broiler output of Thailand.

The C.P. Group began in 1921 as a firm importing vegetable seeds from China. Its success with poultry came in the 1960s when it joined Arbor Acres International in setting up Arbor Acres Thailand to bring parent stock and new technology to Thailand. Arbor Acres Thailand was 51 percent Thai owned. Its second great breakthrough was the establishment in 1973 of a plant to produce frozen chicken for export. Before that date private firms were not allowed to operate a slaughterhouse. It was also the only company to adopt the profit centre concept of management. This enabled it to set up various subsidiaries under competent personnel while its competitors remained family operations. Still 100 percent Thai owned the C.P. Group had in 1982 more than 45 subsidiaries.

Thailand exports frozen chicken parts e.g. boneless breast and leg, to Japan and other countries. It had a cost advantage over United States producers in serving the Japanese market because of lower labour costs in processing boneless meat, which machines could not do well, and shorter transport lines. Its success in exporting to Japan has also been facilitated by Japanese companies such as: Marubeni, Itoman and Co-operative Farm of Japan which handled distribution there.

Some feed ingredients were still imported. However the bulk of the feed consumed and the chickens grown for meat were produced domestically. Thus the industry constituted an important outlet for locally produced grain and materials as well as earning in 1981 around $43 million in foreign exchange.

KENYA COOPERATIVE CREAMERIES

Kenya Cooperative Creameries (KCC) operates on a country wide basis. It was established to process milk surplus to local requirements and export it as butter, cheese and powder. With expanding domestic consumption an increasing proportion of its throughput was marketed as pasteurized milk; exports declined in importance.

KCC was registered as a shareholding company in 1925 and in 1932 as a cooperative. It still operated under both the Company Act and the Cooperative Act. Its membership comprised some 4,000 individual milk producers and 300 dairy cooperatives. It also bought from non-members. Ninety to ninety-five percent of all dairy products manufactured on an industrial basis in Kenya came from KCC. Table 2.15 summarizes its operations for the years 1974 and 1975.

Table 2.15 Summary operations: Kenya Cooperative Creameries, 1974 and 1975

	1975	1974
	(. .litres millions . .)	
Throughput		
Milk and butter fat (as milk equivalent)	224	255
Products sold		
Whole milk	144	143
	(. . . . . tons . . . .)	
Cheese	526	787
Milk powder	5,024	6,417
Condensed and evaporated milk	2,618	3,927
Butter	3,593	4,188
Ghee	433	691
Casein	1	21

KCC received milk and some cream at eight plants organised to specialize in different end products. Supplies arrived by road or rail directly from large farmers (40 to 50 percent) and through dairy cooperatives from smaller producers.

Payment for milk was according to quality, with a bonus for dry season deliveries. It was made monthly 45 days after the month of delivery. Farmers were responsible for collection charges. Shareholder members contributed around one cent per kg. as co-operative dues. Non members paid a 10 percent service charge. The importance of KCC was that it offered an assured outlet. Where district unions could market their milk fresh locally, returns to farmers were much higher - 55 cents per litre for example, as against 40 cents from KCC for first grade milk with 3.5 per-cent butter fat.

Pasteurized milk and milk products were distributed by KCC through sales depots to retailers. It had its own delivery service and also used private contractors. Whole milk for direct consumption went out in tetrapak cartons and to bulk users in larger containers.

KCC had a staff of 1,500. Creamery supervisors were Kenyan Africans with diplomas in dairy technology obtained at Egerton College in Kenya, or overseas. Management was on commercial lines with policy determined by an elected board of directors and implemented by an appointed managing director.

Summary income and expenditure accounts and balance sheets for the years 1974 and 1975 are presented as Tables 2.16 and 2.17. These were good years with sales of whole milk, that brings the highest returns, at record levels. The deficit brought forward from previous years was reduced from nearly $2 million to $232,000 in 1974. In 1975 it was cleared off entirely and $373,000 paid out as dividend, covering also the dividend missed in 1974.

Milk intake into KCC processing plants expanded steadily through the 1970s - to a peak of 273 million litres in 1977/78. It then tended downwards. This was attributed to:

a) an abrupt increase in demand for liquid milk to serve a school milk programme;
b) adverse location of plants in relation to areas of high milk production potential and to school milk distribution requirements, resulting in high transport costs;
c) a government price controls system which KCC had to observe while producers and small producers' cooperatives retailed milk at

Table 2.16 Income and expenditure accounts: Kenya Cooperative Creameries, 1974 and 1975

	1975	1974
	(. $ thousands .)	
Income		
Sales of milk and products	42,382	41,746
Less payments to suppliers	23,942	25,361
Gross margin	18,440	16,385
Expenditure		
Factory costs	11,694	9,379
Sales and distribution	4,129	3,989
Administration	1,436	1,140
	17,259	14,508
Surplus	1,181	1,877
Taxes	572	130
Surplus/deficit brought forward	(233)	(1,980)
Available for distribution	376	(233)
Dividend payments	373	-
Surplus/deficit carried forward	3	(233)

substantially higher prices.

Capacity utilization in KCC's plants was relatively low, perhaps only 50 percent during seasons of low supply. To match changing patterns of production and demand a major rationalization of processing and transport arrangements was required.

While KCC did not offer the kind of assistance to producers furnished in some government sponsored milk projects, it provided valued continuity in processing and marketing services. This was reflected in growing numbers of new members.

Table 2.17 Balance sheet:
Kenya Cooperative Creameries, 1974 and 1975

	1975	1974
	(. . $ thousands . .)	
Liabilities		
Share capital	3,234	3,225
Reserves	2,922	2,351
Long term loans	4,423	2,556
Total liabilities	10,579	8,132
Assets		
Fixed assets	7,730	6,165
Investments	(3)	52
Current assets		
Supplies	4,420	2,196
Stocks	934	1,595
Debtors	2,852	4,050
Cash	1,036	99
Total	9,242	7,940
Less current liabilities		
Due to members	3,606	2,365
Other creditors	2,200	1,671
Taxation	581	178
Bank overdraft	-	1,811
Total	6,387	6,025
Net current assets	2,852	1,915
Total assets	10,579	8,132

ANAND DAIRY COOPERATIVE UNION, INDIA

A classic example of integrating small farmers into a well coordinated processing and marketing system is the Anand cooperative development. It started in 1945 with government entering into an agreement with a private firm to supply milk to Bombay. The arrangement was highly satisfactory to all concerned - except the farmers. The government found it profitable; the firm kept a good margin; milk contractors took the biggest cut. After a 15 day protest milk strike, the Kaira District Cooperative Milk Products Ltd, Anand, was established and formally registered in 1946. It began with a handful of farmers producing 250 litres per day. By 1948, 432 producers had joined the village societies and the Union handled 5,000 litres per day. In 1955, the cooperative's first modern dairy plant was in operation. By 1974, the Anand complex had a milk throughput of 750,000 litres per day and produced butter, milk powder, and other dairy products.

The basis of the Anand pattern was the village cooperative society of primary producers. A milk producer became a member by paying an entrance fee and buying a share. He also undertook to sell his milk only to the Society. Members elected a committee and a chairman, all honorary workers. The paid staff included a secretary, milk collector, butter fat tester, clerk, accountant, and inseminator. Farmers, or their wives or children, took the milk to the society collection centre twice a day. There it was measured and tested. When the farmer delivered his evening milk, he received payment in cash for the morning delivery. In addition to the daily payment on the basis of quantity and quality, the farmers also received a yearly bonus based on the society's financial surplus.

The Union purchased all the milk from its societies, processed it and provided the societies with back-up and production enhancement service. This included artificial insemination centres and services, mobile veterinary units, supply of feed concentrates, and training facilities for society and union staff. The

Union had a nineteen member board of directors representative of its member societies.

Inspired leadership had a vital role in the early stages. It has continued with V. Kurien who was associated with the establishment of the National Dairy Development Board to extend the Anand system to supply other big cities. Known as 'Operation Flood' this programme has been donated large quantities of skimmed milk powder and butter oil for reconstitution as liquid milk. The proceeds of its sale were used to develop and equip with processing and transport facilities further village level societies and district unions. Cooperative milk procurement in India reached 3.9 million litres per day in 1981. This represented earnings of nearly $1.0 million per day for 20 million milk producing families.

Reservations on the later phases of this development were:

1. The cross bred cows introduced to increase yields were less adapted to feeding on crop residues and grain by-products than the traditional buffalo. They competed with humans for their food supply.
2. One third of the aid generated funds were spent on a milk processing and marketing infrastructure considered more appropriate to advanced dairying countries than to the conditions prevailing in India.

On the other hand, Operation Flood certainly brought about a change in the access that dairy farmers had to production technology. Although veterinary services were available through government services or private operators prior to the start of the programme, access was not easy. Operation Flood widened and extended such services through the village cooperatives. Demonstration farms were established to enable milk producers to witness modern approaches at first hand. State development corporations set up a team in each district which worked alongside veterinarians, dairy technologists and extension staff to start cooperatives, organise technical input services, train co-operative personnel and assist in milk procurement. Quality control and value analysis of milk became a village level technology. The organising of cooperatives brought managerial skills to the villages. Incomes were raised in the rural areas concerned and impetus was given to the indigenous manufacture of dairy equipment.

In addition to its economic and technical contributions, the Anand pattern has had an important impact on Indian rural society. The democratic process of election of cooperative leaders helped to break down social and economic divisions. The daily payment often collected by women, enhanced their status and income; it has also provided an income to some landless families who were able to keep cows.

Staff employed by the societies were frequently paid incentive bonuses. An inseminator received an additional fee for each cow got in calf; others were paid a bonus according to the quality of milk delivered to the Union. This creates an interest in the success of the cooperative, an attitude often lacking in personnel employed by cooperatives elsewhere.

CHITALI'S DAIRY, PUNE

For over 30 years Chitali's has been collecting, pasteurizing and transporting milk to Pune from a cluster of villages around Bhilwadi, a small town some 250 km. to the south west. To cope with fluctuations in supply they developed a market for milk based sweets and other products. In 1979 their sales income was made up of $2 million from fresh milk and $4 million from milk based products.

This enterprise provided an outlet for 6,000 milk producers in 400 villages too distant from Pune to sell there themselves. It delivered daily 15,000 litres of buffalo milk in one litre polythene packs to 8,000 households. People buying directly at its seven branches took another 5,000 litres. There were also two dairy product retail shops.

In 1930 Bahashaib Chitali was a small town milk merchant, producing and selling milk in a district about 40 km. from Bhiwadi. By 1946 he was already delivering 500 litres of milk from his own cattle and another 2,000 litres from farmers in the neighbourhood. When he found it difficult to sell all his milk locally he decided to investigate the market in Pune.

In 1980 he had three sons managing the collection, pasteurization, packaging and transporting to Pune of 20,000 to 25,000 litres a day. Three other sons ran the milk distribution and dairy product sales business in Pune. They maintained high standards of quality and cleanliness. Every stage of the operation was closely supervised by a member of the family. In spite of their growing wealth they all worked longer hours than any of their employees.

Chitali's Dairy was in competition with a government milk marketing agency and the Koyna Milk Cooperative. Comparison of their operating costs showed that farmers supplying Chitali received the highest proportion of the final retail price of the milk. At the same time Chitali's business was twice as profitable as the government dairy and almost three times as profitable as the cooperative. Its dairy operated at 97 percent capacity, the government dairy at 83 percent and the cooperative at 76 percent.

Chitali's agents collected milk from farmers, noted the fat content in their pass books and held the milk in ice cooled containers until Chitali's truck passed to collect the milk and deliver it to the pasteurization plant. There were 40 collection agents. Farmers were paid every week on a fat content basis.

To ensure a steady supply of milk Chitali's evolved an animal health care system free of cost for their farmers; they provided artificial insemination facilities and could arrange for cattle feed to be delivered direct to the farm. Their prices were competitive and their dealings fair. The dairy also guaranteed bank loans to farmers for the purchase of buffaloes; at the time of the study there were 1,000 of these loans covering the purchase of 6,000 buffaloes.

The dairy also contributed to the cost of local schooling. However, the main benefit to the producer community was the assured market at fair prices for their milk. The average farmer made a profit of 50 cents per buffalo per day.

At the retail end of the business milk was received fresh every day. If a customer bought some that was sour it was replaced without question. As a result some customers had been buying from Chitali's for two generations. A flourishing business in sweets and dairy products ensured that all surplus milk was used profitably.

This enterprise constituted an effective link between large numbers of milk producers and consumers using processing techniques to balance supplies with demand. It provided veterinary services, fodder supplies and loan guarantees constituting practical assistance to producers in improving their operations. It did this at a profit in competition with a government enterprise that was subsidized and a cooperative that paid no taxes.

DAIRY INDUSTRIES ENTERPRISE, BOLIVIA

This enterprise was set up in 1972 as an offshoot of the Bolivian Development Corporation. Its goal was to implement a national dairy development plan with the following objectives:

a) promote milk production to the benefit of the participating farmers;
b) establish the necessary industrial infrastructure;
c) promote per capita consumption of milk and dairy products to improve nutrition; and
d) substitute for imports.

Its achievements have been substantial. Milk purchases by the five plants it took over rose from three million litres in 1961 to 27 million in 1981.

From the beginning the enterprise set out to help

farmers, especially small farmers, by providing cattle, artificial insemination services, forage seeds, concentrated feeds, small equipment, veterinary and technical assistance as well as regular collection of their milk. These services were provided in kind with compensation by deductions from payments for the milk. This system has the advantage of avoiding the formalities and guarantees needed for bank loans. It also eliminated the risk of cash being used for other purposes. Finally, the farmers were committed to sell their milk to the plant. They seemed to be satisfied with this system, as evidenced by the considerable increase in numbers participating between 1975 and 1978.

For many small farmers milk production for sale was new. The Dairy Industries Enterprise set up dairy development modules in partnership with them to supply inputs and services. The enterprise recruited six agronomist/animal husbandry technicians to advise farmers in collaboration with five veterinarians. Training courses were organised.

Payment was according to fat content with a lactometer reading to detect adulteration. Acid milk was rejected, but there was no premium for clean milk production, and in some areas the bonus for higher fat content did not fully reflect the market value of the butter fat. Producers were paid once a month but there was generally a fortnightly advance payment.

Pasteurized milk accounted for 50 percent of sales. Other products were flavoured milk, cream, ice cream, milk based sweets, butter and cheese. The plants operated at a profit, as required by the enabling legislation and this was reinvested in dairy development.

These plants were handicapped, however, by underutilization. This was attributed to over optimistic projections of supply and demand and the availability of cheap external funds at the time of planning. Lack of protection against imports of milk powder and condensed or evaporated milk compounded the problem. Other handicaps in competing with private vendors and manufacturers were a centralized inflexible marketing and price policy and high collection and distribution costs. Retail prices of milk often lagged behind inflation because of government controls. While per caput consumption of milk products in Bolivia rose during this period, low income consumers tended to prefer powdered or evaporated milk for lack of refrigeration in the home and because liquid milk distribution facilities in the areas where they lived were inadequate.

Overall, the operations of this enterprise, which received substantial World Food Programme assistance during this period, were geared to bringing small farmers into the dairy economy. In order that milk collection and product marketing could be adapted more closely to local conditions, management of some of the plants was decentralized.

DRIED FISH, INDONESIA

Over half of Indonesia's total fish landings in the early 1980s were converted into cured products so that they could reach consumers in acceptable condition. During the period 1976 to 1980, the percentage of fish caught in Indonesia and used for dried, salted or smoked products rose from 36 percent to 58 percent, with the wet weight input increasing from 464,000 to 714,000 tons. This is significant because it is the reverse of the trend found in most countries in the region.

The processing methods used in Indonesia were similar to those employed in other South East Asian countries with the exception that Java consumers used a lot of "pindang", fish boiled in a heavy salt solution. Catching and processing technology was traditional with the exception of jelly fish drying which was developed with Japanese assistance.

The dried fish market handled about 440,000 tons annually. Dried fish products were brought by boat to Jakarta, Geribon and Surabaya and thence distributed to various Java markets by truck. Exports were mainly dried jellyfish and salted fish roe shipped to

Japan. Total export volume in 1979 was 3,370 tons valued at $7.8 million.

The dried fish production/processing/marketing system was made up of the following:

1. Producer: The fisherman who landed his catch at auction centres or might process part of his catch into dried products.
2. Dealer-processor: He would buy dried fish products from fishermen and also process fish bought from landings or at auctions. He might sell some of his products to local agents or retailers, but his principal market was the urban wholesaler.
3. Wholesaler: Usually he was located in a major urban centre and received dried fish consignments from other areas. From dealer-processors these might amount to 400 tons at one time. The fish arrived sorted and prepared for auction to agents, other wholesalers or large retailers.
4. Agent: He bought large quantities of dried fish from wholesalers and distributed them to retailers. He might also operate one or more retail stalls.
5. Retailer: Normally he had a stall at a market; he received dried product from the agent, prepared it for display and sold to the public.

The heart of this system was credit in which the dealer-processor appeared to have the most at risk because he had to pay cash for his raw product and occasionally advanced money to fishermen. The wholesaler ordered on the basis of a 50 percent advance payment for each shipment. The balance was paid after the fish had been received and the quantity and quality had been checked. This might take as long as one month. The wholesaler auctioned to agents in lots of 300 kg. Credit was extended to the agents for two to four weeks. They, in turn, allowed one to two weeks credit to retailers, sometimes giving them stocks on consignment.

The principals in the market chain were the dealers, wholesalers and agents who handled large quantities. They had a sizeable amount of investment at risk. This section of the chain was invariably composed of ethnic Chinese who had done business with the same families for years or even generations. Working with them the middle link, or wholesaler, had less credit risk, whereas dealers and agents incurred as much as five percent credit losses. Obviously, they allowed for this in their operating margins.

Operating margins through the chain can be estimated only broadly. The processor expects to clear about 10 percent on his overall outlay which would include purchase of raw product, salt, plant labour and transport to market. The wholesaler makes a margin on his costs plus an auction fee which varies from one to five percent. A one percent government tax is applied to dried fish auction proceeds. Agents probably operate on a 10 to 25 percent gross margin. The same general percentage would apply to retailers.

This is a long established system in which most of the technologies used are traditional. The critical link seems to be finance of which the processor carried the brunt.

LOBSTER TAILS, BRAZIL

Lobster were fished in Brazil in the 1970s by a range of enterprises. There were four with fleets of 40 boats 6 to 15 metres long. Others operated with one or two boats only. The larger boats had their own freezing equipment. Others packed their catch in ice until it could be frozen on shore.

Marketing was highly fragmented. From the USA, the main market for frozen lobster tails, some 15 importers travelled frequently to Brazil to buy directly from fishermen or via local wholesalers. They then competed against each other to sell to distributors and restaurants in the USA.

In 1976 the Government of Brazil organised Interbras as a state trading company to assist in developing the exports it needed to pay for its imports of oil. The initiative for Interbras' entry into lobster exports came from Internor, its North American sales agency subsidiary. The manager of Internor saw the potential advantages of coordinating imports of lobster tails and made direct contacts with captains of fishing boats. He met scepticism initially and outright opposition from those who already had good relationships with particular exporters. Interbras began its programme in 1979 with some 50 percent of the skippers. Those joining subsequently raised the percentage to 90.

Interbras began by offering a price higher than that obtained by most of the fishermen, an offer that was good for a three months period. In the USA it also offered a firm sales price for three months to distributors and restaurant owners selected to avoid market overlapping. It took a considerable risk since traditionally the spot price had fluctuated with variations in supply. However, frozen lobster tails can be held in storage for three months, an important advantage in maintaining firm prices. A further advantage to producers was that they shared in any profits Interbras obtained from sales of lobsters above the prices needed to cover its marketing margin.

Interbras gave Brazilian lobsters a market identity which they had lacked before. This had been a disadvantage in competing with lobsters marketed by the South African Seafood Association. Before the Brasmar label was adopted the various producers and exporters had used their own names, often this confused North American buyers. Promotion in the trade press was also facilitated by use of a single brand. A further innovation was rigorous quality control and packing by uniform size. This was specifically adapted to the needs of restaurant buyers interested in "portion control". This practice was carried back to the original producers and packers. Interbras also stressed immediate cooling of lobster to avoid the 'black spot' defect.

Included in the Interbras' operating margin were the interest charges for financing cash payments to producers and all marketing costs until payment was received from US distributors. Holding stocks in public cold storage warehouses in major distribution centres was part of this operation. Credit for fishing boats, etc., was available under government programmes.

In the early 1980s the lobster marketing operation of Interbras involved a staff of 15. Its collective export marketing was returning to Brazil $6 million more annually than the former system.

FROZEN SHRIMPS, ECUADOR

Exports of shrimps from Ecuador increased from about 4,000 tons in 1977 to 12,000 tons in 1981. In 1982, exports amounted to $148 million primarily to the USA. Foreign exchange earnings from shrimps exceeded those of bananas, coffee and cocoa and were the second largest to oil.

The following annual production estimates (tons) show the rapidly increasing share of cultivated shrimps in the total:

	Cultivated	Sea caught	Total
1979	4,698	7,787	12,485
1983	29,100	7,500	36,600

The area of water under shrimp cultivation in 1983 was 45,500 hectares. The majority of farms were less than 100 hectares and operated extensively. Juveniles and post-larvae taken from the wild were put into grow out ponds with a stocking density of less than 15,000 per ha. with no supplementary feeding and no water exchange.

Increasing capital costs of land and preparation together with technological advances led to semi intensive cultures. These involved continuous water exchange, nursery and grow out units and a carefully balanced feeding regime. Stocking rates were 80 to

100 thousand per ha. Yields averaged 2.0 to 2.2 tons per ha. The rate of return on new investment was around 40 percent. Technicians were brought in from other countries to provide expertise in farm and hatchery management. By 1985, however, production was declining because natural reproduction had run down through over fishing. Hatchery technology had still to be developed. Inadequate supplies of feed were also a constraint in 1983, since overcome by incentive prices for maize the main ingredient.

In parallel with the expansion of shrimp farming has been a rapid growth in shrimp processing companies and exporters. In 1985 there were 65 processing plants. The larger and longer established both owned vessels and had others fishing for them under contract. Most of the farms were independently owned, but the larger, more sophisticated farms tended to be owned by the companies which had their own processing plants. Quality control was supervised by the Istituto Nacional de Pesca.

Exports to the USA went mainly to brokers on consignment. The exporter was advanced 60 to 80 percent of the expected price by letter of credit. The balance was remitted after the product was sold, less brokers' costs and commission.

ALLANA FROZEN FISH, BOMBAY

Allana was founded in 1865 by a Moslem from Gujarat. While still family run, it had become India's biggest food trading company in the 1980s. It handled a range of lines and had developed the export of Halal slaughtered meat to Middle East markets. Its entry into fish marketing originated with an enquiry from an importer in Kuwait as to whether it could supply pomfret the preferred fish in that market. This was in 1976. In 1982 Allana exported 6,000 tons of frozen fish to the value of $14 million.

The sales strategy adopted was one designed specifically for the discriminating Kuwait market, to offer whole, fresh frozen pomfret, individually wrapped in a polythene container; each wrapper was stamped with the date of freezing. They were shipped to the wholesaler in 20 kg. net cartons. For the 200,000 expatriate Indians living in Kuwait and Saudi Arabia Allana offered lower priced mackerel in one kg. cartons packed for supermarket distribution.

To attain the quality of fish desired and break into established supply channels Allana began by offering fishermen 20 cents per kg. over the going price. Later, Allana's buyers waited until local consumers' orders were filled before making their bids. The company's policy was to buy the best fish then available and to begin quality control immediately by reducing the time and distance between landing and freezing to a minimum. It established freezing plants conveniently near to key fish landing points. Workers washed and cleaned each fish, checked it for quality then inserted it in its polythene bag, prior to plate freezing and pressing under 140 kg. of hydraulic pressure. Between processing plant and port of shipment, Allana deployed a fleet of 35 company owned and 15 leased insulated 12 ton trucks. For shipments to the Gulf the company had a 2,300 ton refrigerated ship Al Gilani, India's first fully refrigerated vessel. It had 4,000 cubic metres of storage space in 15 compartments each with its own temperature control.

Allana came to the fishing industry with a promising export market outlet with which it was already familiar, with marketing expertise, and with capital. It acquired processing technology and set up plants at favourable supply points. Its approach to the fishermen was to pay for produce adapted to its market, counting on steadily expanding supplies for which its purchases contributed an incentive. Finance for fishing operations has remained difficult with pressure mounting for the establishment of a fisheries bank.

NOUADHIBOU, MAURITANIA

The Islamic Republic of Mauritania is mostly desert. However, with general recognition of the principle of 200 mile exclusive economic zones, it acquired fishery resources to match those of Canada or Norway. Traditionally, domestic demand was met by canoe fishermen scattered along its coastline. Since 1979, an offshore catch of 300,000 to 400,000 tons per year has been processed and marketed through enterprises based in Mauritania. Formerly much of this was taken by foreign vessels and marketed through Las Palmas. Nouadhibou became the centre of fishing in Mauritanian waters. A port berthing 15 to 20 vessels was constructed; ice making plants, processing and freezing establishments, and many joint ventures with international interests appeared. There was a private capital inflow of over $60 million and a parallel government investment of $21 million in fisheries infrastructure. Exports of fishery products reached 286,000 tons in 1983 bringing in $148 million of foreign exchange. One quarter of these exports went to other African countries. The major markets were Japan and Russia. Exports were mainly cleaned frozen fish, canned fish and fish meal (90,000 tons). In total, fisheries products amounted to 19 percent of the gross national product.

Sales from artisanal fishing were free but, to control over the side sales since 1984 all off shore fish had to be sold through a state monopoly Societè Mauritanienne de Commercialisation de Poisson (SMCP). It operated on a non profit basis according to commercial principles. Offers received over the past ten days were reviewed by a commission. On the basis of these offers prices were set for the next ten days. Two representatives of the fishing companies' association FIAP sat on this commission. SMCP had an office in Las Palmas serving European markets and another in Tokyo. It was operating in the black.

The industry association FIAP had 61 member companies and was an important link with the government. It was concerned as how best to integrate into the industry the artisanal fishermen. While their sales on domestic markets proceeded separately, they also supplied high value products such as live lobster to the export companies. Mauritanians were also gaining experience in large scale fishing and processing operations through joint capital and technology sharing ventures.

FISHMEAL, PERU

During the 1950s and 1960s fishmeal was the most dynamic export industry in Peru. Foreign exchange earnings from fishmeal increased a hundredfold from 1955 to 1964: it was Peru's leading earner of foreign currency. In 1964, 1.5 million tons were produced and 1.4 million tons exported.

In Peru, fishmeal was manufactured almost entirely from anchovies, which abounded the 1,400 mile coast. They were cooked, pressed, dried, and milled into a dry meal. The liquid extracted contained fish oil an important by-product, and solids which were recovered to improve the yield of meal. The meal was held in bulk or in 50 kg. jute or treated paper sacks.

Fishing was not uniformly good along the coast; the fortunes of any one zone varied from season to season or from month to month. Because of the risk element in the availability of fish the trend was for anchovy processors to seek greater control over their raw material by operating their own fleets. While most of the fishing was originally by independent boat owners who sold their catch to the factories, by 1966 70 percent of the fish were landed by boats belonging to fishmeal plants. The consensus among plant owners was that they would prefer not to undertake the additional organisation and higher fixed costs involved in operating their own fleets. However, the independent fishermen proved unreliable; they did not always keep to their contracts and quickly deserted areas of poor fishing. Thus a plant maintained a fleet to guard against low capacity utilization or a complete shut down if the fishing in its area declined.

This risk also led to processors operating plants in several zones.

In 1968 the largest producer Luis Banchero, a Peruvian who entered the industry in its early stage had eight plants with an output of 254,000 tons of meal, 13 percent of the total for the country. The next two processors were International Proteins Corporation of the United States, with five plants, and the Peruvian partners, Madueno and Elguera, with four plants each producing about six percent of the total.

Initially processors marketed independently, responding to bids from European and North American buyers for rapidly growing poultry and livestock feed industries. In the mid sixties there was a move towards stabilisation of the market via the allocation of quotas by a National Fishing Society. It represented Peru in an international Fishmeal Exporters' Organisation. By 1968 the processors had formed the Fishing Consortium of Peru with 62 members representing 59 percent of the supply and three lesser groupings each of which marketed all its fishmeal through a single trading company.

The first plant devoted entirely to the reduction of whole anchovy was set up in 1950 in Chimbote, with used equipment imported from California. Called Pesquera Chimu, it was a joint venture of the Wilbur-Ellis Company of San Francisco, which had been interested in Peruvian fishing before the war, and Manuel Elguera, who had been assistant superintendent for one of the first commercial fishing ventures in Peru and later became the third largest processor. The availability of nylon nets was part of the original breakthrough. These came from Japan.

About this time the commercial banks began to take an active interest in the industry. Processors could obtain finance easily though interest rates were high. A national equipment industry came into existence to supply reduction plants and fishing fleets. The need to move whole anchovy into and out of boats in large quantities led to the development of a specialized fish pump by the Peruvian company, Hidrostal; it used a helical impeller to avoid cutting the fish in transit.

Technology spread by rapid copying under the incentive of high profit prospects. In 1963, 36 new plants were constructed, and 453 boats were added to the fishing fleet. All the boats and an increasing proportion of the processing equipment were supplied by Peruvian manufacturers.

Access to foreign capital was of critical importance in the development of the fishmeal industry, but it did not dominate it. In the later stages of expansion, particularly, it came in indirectly by bank borrowing of dollars to back their local currency loans. This left control in Peruvian hands. About one quarter of the equity capital was supplied by foreign firms and individuals. Over 100 of the leading entrepreneurs were Peruvian. Only 13 of the 125 fishmeal firms were controlled by foreigners while another eight had foreign participation.

Through the development phase the Peruvian Government had only a minor role. It came in with the call for market stabilisation but left the allocation of quotas to the industry's own society. Later, with supplies declining because of over fishing it established periodic 'vedas' of one to three months when no fishing was allowed. After a further decline in supplies and various amalgamations and bankruptcies a radical government nationalized the whole industry. In the 1980s the state enterprise Pesca Peru was trying to shed 20,000 personnel in an effort to reduce its overheads and balance its books.

STRAITS FISH MEAL, MALAYSIA

This fish milling enterprise based at Bagan in Selangor state began as a family business in 1969. It became an incorporated company with a paid up capital of $208,000. While the management was ethnically Chinese, the firm had a Malaysian chairman and other Malaysian shareholders. Together they had contributed ten percent of the capital. The company bought fish not meant for human consumption, cooked, dried and ground it, then packed it in plastic bags for sale to poultry feed companies. Poultry was a growing industry in Malaysia and constituted a good market. In 1984 it imported from Thailand a new set of drying ovens incorporating the most recent technology. The mill operated on a 24 hour basis with 60 employees working in three shifts. Average monthly out turn was 400 tons of meal.

The bulk of the fish purchased by the mill was supplied by fishermen living in villages nearby and along the Straits of Malacca. To be sure of supplies the mill advanced $1,000 to $2,000 per boat before it put out to sea. It financed some 200 boats involving a total credit outlay at one time of over $300,000. These advances were cleared against payments for the fish brought in. No significant amount outstanding on this account appears in the company balance sheet at the end of the year. (See Table 2.18) However, sales of meal to feed companies were made on a one to two months credit basis. In addition, the mill normally held stocks equivalent to two months sales so that it could deliver promptly when orders came in. To meet these financing needs the firm acquired its new dryer costing $146,000 through a local leasing company. It had a credit line of $200,000 from the Agricultural Bank. The collateral for this was its industrial land title and the "joint and several" guarantees of all the directors. The mill had a long standing clientele of nine feed mixing enterprises. Their market was growing and the mill's product had a good reputation for quality.

On the basis of current prices for fish unwanted for human consumption $66 per ton and of fishmeal $482 per ton the mill's income/expenditure position was as follows:

	$ per ton of meal
Cost of raw fish	261
Overhead and miscellaneous expenses	103
Fuel (sawdust, wood, diesel oil)	36
Total costs	400
Sale of meal	482
Profit margin	82

Table 2.18 Balance sheet: Straits Fish Meal, years ending 31 December 1982 and 1979

		1982	1979
		(. . .$ thousands . . .)	
Assets			
Fixed		188	168
Share in subsidiary co.		13	-
Current:	stock in hand	85	40
	trade debtors	250	143
	other debtors	3	30
	advances	1	2
	cash	1	8
Total assets		541	391
Liabilities			
Share capital		208	208
Capital reserve		17	17
Retained profit		104	75
Hire purchase credit		12	-
Trade creditors		56	19
Provision for taxation		20	-
Bank overdraft		124	72
Total liabilities		541	391

Operating at capacity, the price margin between raw fish and processed meal would return to the firm a net income of $33,000 per month. With this prospect the firm was seeking additional funds to expand its fish purchasing base. Such returns were not always there, of course. In 1980 the firm's net profit for the year was only $3,300. In 1979 it made a loss.

The firm's profit and loss figures are shown for years 1982 and 1979 in Table 2.19. It will be noted that since this is a limited liability company, the directors' emoluments are shown as a cost. They also benefit through their participation in the retained profit. This is used as a reserve. It absorbed the loss incurred in 1979 and was built up further in 1982.

Table 2.19 Profit and loss account: Straits Fish Meal, 1982 and 1979

		1982		1979
	(. . . $ thousands . . .)			
Sales		1,659		567
Closing stock	186		36	
Opening stock	170		29	
Increase in stock		18		7
Total		1,666		574
Costs				
Raw material and processing		1,476		494
Transport charges		63		14
Packing costs		10		4
Advertisement		1		1
Depreciation		24		15
Office and administration		40		49
Directors' emoluments		31		8
Total costs		1,645		585
Profit		21		(11)
Retained profit brought forward		92		85
Retained profit carried forward		104		74

In financing Straits Fish Meal the Agricultural Bank took account of the benefits it brought to fishermen. It guaranteed them an outlet for their fish. It provided finance that presumably they could not obtain on more favourable terms from another source. In regard to consumers it furnished an essential protein ingredient for poultry feed that could not be obtained more cheaply elsewhere and so helped make eggs and meat poultry available at lower cost.

3 Alternative enterprises for processing

A reliable raw material supply and effective marketing of the product are crucial factors in the success of a processing enterprise. With them must go:

a) selection of suitable processing equipment and its maintenance in good working order;
b) assuring that the product meets required quality standards;
c) timely procurement of other processing ingredients, containers, fuel, services, etc.;
d) recruitment and supervision of suitable technical, office and processing plant staff;
e) assuring adequate and timely finance, payments to suppliers and collections from purchasers;
f) operation at the lowest practicable cost consistent with efficient performance;

PRIVATE ENTERPRISE; TRANSNATIONALS; COOPERATIVES; PARASTATALS: ADVANTAGES AND LIMITATIONS

In principle all kinds of processing enterprises can aspire to meet the above requirements. In practice experience suggests that some are better for some purposes and under some conditions than others. In this chapter we shall compare the ability of enterprises in alternative forms of ownership to integrate these management components. Account will be taken of different product characteristics, raw material production environments and whether sales are directed primarily to export or domestic markets. While the case studies in Chapter 2 include many of the most successful processing enterprises in the developing world it is recognized that they do not constitute a sample that can be a basis for general conclusions. Other relevant studies and information will be taken into consideration. The management of processing enterprises in general will be taken up again in Chapter 5.

Alternative enterprises for processing fall under four broad headings:

1. domestic private enterprise;
2. transnational enterprise alone, or in joint venture with local interests;
3. cooperatives or farmers' associations;
4. state corporations and other parastatal bodies.

The private business enterprise is one in which the capital is owned directly by the manager of the enterprise, by him in partnership with others, or by private investors who have acquired shares in a company. Such enterprises can be set up by an informal personal or joint decision. Those with shareholders whose financial responsibility is limited to the shares they have taken up have to be registered under legislation governing the establishment or limited liability companies. Generally this can be done on payment of a small fee.

A transnational enterprise is one that operates in one or more countries foreign to that of its headquarters. Conspicuous are the large joint stock corporations that have established international brands. However, a transnational joint venture can also be small scale - collaboration between brothers located in different countries, for example.

The characteristic feature of a cooperative is that it is owned by those who use its services; they are entitled to share in any profits it makes. It is also managed democratically by its owner members. A committee elected on a one member one vote basis directs its affairs. It can appoint a manager but he remains responsible to the committee.

State corporations are enterprises set up by government with government capital and a directing committee appointed by government. Usually they are autonomous in their day to day operations, but are directly responsive to government instructions. Grouped with them as parastatals are agricultural marketing boards and area development and settlement management authorities assigned legal powers by government to undertake commercial operations on their own account.

On the basis of a review (Abbott, 1987) of relative success in food and agricultural marketing in the developing world, the following hypotheses are presented for enterprises undertaking the processing of agricultural and fisheries products.

Domestic private enterprise

Private business enterprises have shown themselves well suited to:

a) Take advantage of and exploit unforeseen opportunities and follow up new ideas.
b) Start up and go a long way with very little capital. Private traders tend to be economical, even parsimonious, in their personal expenditure, very careful in their business outlays.

c) Operate at very low cost. Only those staff are employed who make a positive contribution to the enterprise. Full use is made of family labour available at no cost.
d) Because decision making is concentrated these enterprises tend to show ready initiative and quick response to changing situations.
e) Family ties and kinship linkages can often be used to extend the marketing operation with high confidence and low risk.
f) A continuing sanction against inefficiency in a private enterprise is that, unless there are barriers to the entry of new firms, it will lose customers and go out of business.

Areas of agricultural and fisheries marketing where private enterprises tend to perform better than others include:

a) Perishable products. Variability in quality, a tendency to deteriorate quickly if not held in special storage and sharp changes in price in response to variable supply call for rapid responses on the part of the enterprises handling such products.
b) Livestock, meat and fish. The need for judgement in appraising quality and value and for care in handling to avoid losses gives an edge to direct decision-making. The predominance of private enterprise in the marketing of livestock and meat also reflects a reluctance of many people to come close to the realities of this trade. These considerations also apply to fish.
c) Combined assembly of produce with provision of other rural services. When the quantities supplied by each customer are small and varying, considerable local knowledge, patience and willingness to provide a number of services over a wide range of hours and locations are needed.

d) New and highly specialized activities. Characteristically these are the outcome of an individual initiative, not a planned development by a committee or a government department.

Characteristic limitations of indigenous private marketing structures are:

a) constraints on access to capital;
b) temptations to collude on prices where traders' interests coincide;
c) varying management capacity.

Transnationals and joint ventures

Potential contributions of transnational enterprises are:

a) Finance. Generally they are in a position to mobilize capital from the lowest cost sources. It can be brought in as equipment, improved seeds, strategic supplies and skilled management and technology, for which foreign exchange would be needed in any event.
b) Applied technology. Developing countries face the risk of selecting unsuitable designs and equipment and the problems of putting new plants into operation and maintaining them. Engaging an enterprise with demonstrated experience in applying a desired technology and in a position to keep it up to date is often the safest, and, in the longer run, least expensive way of acquiring it.
c) Management. When qualified management experienced in the handling of specific product lines comes with a transnational it is an immediate advantage. Local personnel can learn from it by working with it.

d) Quality standards and presentation. The transnational experienced in meeting strict standards can help a developing country overcome such barriers to successful export marketing. It can also reduce quality risks to domestic consumers and help adapt domestic agriculture to produce raw materials with the required attributes.
e) Market access. In export sales a close link with an enterprise which has established outlets in major import markets is a great advantage.
f) Brands. These carry great weight with consumers, and the wholesalers and retailers who serve them. An agreement to sell through the owner of an established brand enables the producer to share in the benefits of past outlays on its promotion.

There was a period during the 1970s when developing countries became afraid of the transnationals because of the financial weight they wielded. They saw a political risk in being dominated. They saw the commercial risk of farmers growing for a transnational outlet being left without a market should it decide to withdraw. Since then, much of the steam has gone out of the issue of transnational power:

1. Because of the uncertainties of foreign investment the transnationals have tended to shift from production in a developing country to the sale of technology, management services and marketing. The technical revolution in information transmission has eroded one of their formerly unique advantages, their global communication systems. This has lowered the entry barrier for small and medium sized trading companies.
2. The panorama of transnationals is now much wider - no longer with an aura of neocolonialism. It includes many with their headquarters in other countries including

Japan and in some developing countries.

3. Transnationls are becoming more varied and more flexible including banks, retailing chains, management and consulting firms and training agencies.

Alternatives are available to take the place of particular transnationals. The situation has become one where the government of a developing country can assess the benefits that a transnational investment or collaboration can bring to its economy and bargain over the terms. For such bargaining, and subsequent monitoring of prices and profits, it should have access to the necessary information and skills.

Cooperatives and farmers' associations

The marketing efficiency of a group of farmers is increased by selling together where they can:

a) benefit from economies of scale in the use of transport, processing, storage and other services through increasing the volume of a commodity handled at one time;
b) raise their bargaining power in sales transactions.

Conditions recognized as favouring cooperate marketing are:

a) specialized producing areas distant from their major markets;
b) concentration and specialization of production;
c) homogeneity of production and output for market; and,
d) groups of farmers dependent on one or a few crops for their total income.

Factors favouring successful cooperative marketing are:

a) availability of local leadership and management;
b) a well educated membership;
c) members all belonging to one family grouping i.e. with strong kinship ties or integrated by religion.

In developing countries cooperatives have shown themselves well suited to undertaking the:

1. assembly of fairly standard not very perishable products such as coffee and cotton where rapidity of storage and transport is not critical, for sale to pre-established outlets where the price risk is small;
2. assembly of a product such as milk for a pre-established outlet where the crucial requirement is regularity of collection and procedures for protection of the product and assessment of quality have been standardized.

Cooperatives have difficulties:

1. in mobilizing capital because of the intrinsic requirement that it come from their members on an equal basis, and because they lack collateral to offer as security for loans;
2. in securing adequately qualified management because of their members' reluctance to agree to a remuneration much higher than their own;
3. in optimizing marketing opportunities because of the need for policy endorsement by members lacking marketing experience and expertise.

Nevertheless, where the absence of a cooperative alternative channel could mean that farmers are less well treated in marketing, its maintenance through government support can be justified.

State corporations and other parastatals

These bodies are generally created to set up and run facilities that would not be provided by existing enterprises. They are also used to implement government policies to assist particular economic groups and to undertake monopoly roles judged to be in the public interest.

A practical advantage of such enterprises in recent decades has been their eligibility for external aid. International and bilateral assistance agencies have tended to work with government bodies because they needed official counterparts. Thus 487 agro-industrial components of projects financed by the World Bank/IDA 1972-84 were implemented by parastatals, 14 by cooperatives and seven by private enterprises.

The characteristic limitation of the state enterprise in the performance of commercial functions is that its management and staff rarely have a direct incentive for efficiency. While the autonomous parastatal is certainly better suited to commercial operations than a government department with restrictive financial procedures, many are still tied too closely to civil service salaries and conditions of employment. While ways are found to add staff, many parastatals find it extremely difficult to terminate them. Management capacity must be sufficient to overcome traditional attitudes and competing loyalties of staff. It has still to face the depredations of politicians with their own immediate interests.

Characteristic of many developing countries is the agricultural marketing board. These bodies may be endowed by the government with legal powers to require all producers and handlers of a defined commodity to follow policies it establishes. These can range from regulation of the qualities of produce that can be sold in particular markets to requiring that all produce wholesaled pass through the board. The marketing of major export crops is often assigned to such a board, sometimes via nationalization of foreign owned firms. The government then has direct control of this source of foreign exchange. State corporations and marketing boards also undertake supply and price stabilisation operations for basic food crops. They stand ready to buy from farmers at a pre-announced minimum price. They hold stocks for release to retail outlets when consumer prices reach an unduly high level. They may operate their own mills.

There are commodity marketing situations where parastatals are common and others where, for practical reasons, they have been found less convenient and effective. Supply and price stabilisation agencies are concerned with the major food grains - maize, rice and wheat. Less 'political' grains and pulses, including some often consumed by lower income people, receive less attention because of governments' needs to limit activities that might call for subsidies. Coffee, cocoa and cotton typically sold by standard quality specifications are widely handled by export parastatals. Tea and tobacco requiring direct examination of samples are more often sold by auction. Livestock and meat, perishable fruits and vegetables, and relatively perishable tubers also tend to be left to other forms of enterprise.

Let us now see how far these hypotheses constitute a practicable criteria of the suitability for various tasks of these alternative forms of enterprises as processors.

EXPERIENCE BY COMMODITIES

Rice

While rice is a fairly standard, relatively durable product, rice milling is predominantly in the private sector. Rice growers like Enebor in Nigeria went into milling as a service to nearby consumers. They found it profitable and continued to expand in processing. There is little evidence of transnationals' interest in rice milling. Apparently the scope for adding value has not been enough to attract them.

Rice milling by farmers' cooperatives has been

promoted by governments in various countries where private millers were considered to be taking monopolistic profits. Rice milling cooperatives have rarely, however, achieved a significant market share except where this has been assigned by political decision.

Parastatal rice mills have been set up as part of a government sponsored development programme, typically to serve intensive irrigated production, as in the Niger project in Mali. There was concern that adequate facilities be available when new settlers started production. Scale economies from the construction of large processing units to serve intensive production were anticipated. Some supply and price stabilisation agencies have acquired milling capacity. This facilitated direct injection of rice into consumer markets. However, the quantities handled by such agencies have varied greatly from year to year with a corresponding impact on mill utilization and costs.

Maize

Much the same structures occur in the processing of maize where it is a staple in human consumption. Private enterprise undertook rural milling; it expanded to serve the larger consuming populations of urban centres. In some countries, such as Tanzania and Zambia, the big urban mills have been nationalised and are operated by parastatals. In Kenya, in contrast, some are in influential local private hands. Corn Products Corporation International has gone into maize milling where surpluses to human needs favoured the processing of maize into industrial products for use domestically and for export.

Cassava

Transnationals saw the livestock feed potential in Europe and developed an economical technology for pelleting. They led the export boom in Thailand.

Associated enterprises operate some rural mills and pelleting plants but many are independent.

Feed

Commercial feed mixing is new in the developing countries: it has been a private enterprise initiative. Ratna Feed, a family enterprise was the first in Nepal to foresee the need to promote poultry production enterprises as outlets for its products. It assumed major responsibilities for advising and assisting them in both production and marketing.

Fruit

Large scale processing of fruit in the developing countries is a domain of the transnational. Their goal has been to convert it into a form that can be exported to higher income consumers, able to pay for variety in their diet. The transnational supplies technology, management and access to the export market. The Dabaaga case is one of processing fruit for a relatively small, medium income domestic market. It was started as a family enterprise with very low costs. It is also nigh transnational since the owner came from Goa and looked to India for appropriate technology.

Vegetables

Commercial processing of vegetables has followed a similar pattern. The first enterprises were transnational, as illustrated in tomato canning in Taiwan. However, with the technology widely disseminated and tomato paste now a basic ingredient in many low income diets, the processing of tomatoes has become a characteristically local operation in many developing countries. There are some state corporation and cooperatively owned plants, but private enterprise predominates. Pepper processing in Honduras and asparagus canning in Lesotho were initiated as cooperative and state enterprises respectively, with external assistance. It remains to be seen if they endure.

Flowers, herbs

The processing of marigold heads in Ecuador and of medicinal herbs in India are characteristic of private enterprise's seeing new opportunities and taking them.

Sugar

This has been a domain of the transnationals. They went in to serve metropolitan markets where they had their own outlets. The Japanese Thai case reflects continuation of this link. With declining consumption and protected domestic production in many developed countries, the people of the developing world now constitute the main new market. The trend in the 1970s and 1980s has been for new plants to be government projects, with transnationals engaged as managers and advisers. Booker McConnell in Kenya, Papua New Guinea and Sri Lanka, H.V.A. in Ivory Coast and Schaffer in Sudan follow this pattern. Major inputs of capital are needed for an efficient production and processing unit. Profitability is likely to depend on government pricing. The parastatal seemed the logical enterprise.

Tea

Labour problems when cane had to be cut by hand during a short season contributed to the move of the transnationals out of sugar. With tea, the labour availability issue is less acute because picking goes on through the year. Nevertheless, the Kenya Tea Development Authority with its focus on smallholder production was a far seeing approach in associating the picker in the success of the overall enterprise. It demonstrates how a parastatal with monopoly powers over tea growers in a defined area can bring a range of benefits. Credit goes to the Kenya Government. A parallel project in Tanzania over the same time period did not do so well. It was required to take on the management of nationalized private plantations while already struggling with resource and management problems.

Cotton

This is a major cash crop for small farmers in the developing world. Its marketing impinges on the livelihood of rural families in large numbers. It is a standard durable product. Primary processing, separation out of the seeds and baling the lint into uniform packs, is a basic technology. For these reasons it has often been considered a convenient activity for farmers' cooperatives. They became the assembling agents of a marketing board or state backed company. In Uganda it was recognized in an official enquiry that small Asian traders could carry out the same function at lower cost. It was maintained, nevertheless, that farmers preferred their cooperatives because it was felt that they could be trusted.

Oilseeds

Crushing oilseeds has been a traditional small scale private enterprise activity in areas where they have been the basic source of cooking fats, as in tropical Africa. This holds good also for olive oil production in countries bordering the Mediterranean. Transnationals first entered as buyers of oilseeds for processing into a range of food, soap and other products for metropolitan markets. They then set up plants in developing countries where domestic demand was sufficient to carry the overhead costs of their technology. While local competition has appeared, transnationals still have a major role.

Meat

The processing of livestock into meat and various by-products goes on everywhere as a private enterprise, often by separate communities because the bulk of the people do not want to be involved. To ensure a minimum of sanitation municipalities have provided slaughtering slabs, eventually municipal abattoirs where wholesalers have their animals slaughtered. The first large scale operations were set up by transnationals to obtain low cost meat for export to deficit metropolitan countries. From Africa this was limited to canned meat because the cooking involved gave protection against the transmission of exotic livestock diseases. The higher potential returns from marketing uncooked meat frozen brought in governments. Importing countries required them to guarantee that the meat exported came only from disease free animals. Often only governments could mobilize the funds needed to set up facilities meeting the required standards of hygiene. In parallel, private enterprise has invested in improved facilities to serve specialized markets; witness La Favorita in Ecuador and Indian enterprises serving the oil rich consumers of the Gulf.

Hides, skins

Processing hides and skins is traditionally a specialized private enterprise. In various developing countries commercial tanneries were set up by Bata, the transnational footwear retailer. It had its own outlet for their products.

Poultry

The processing of meat poultry was undertaken until

relatively recently by individual producers and retailers. Close integration with production and input supply is now widespread. Direction and financial responsibility are generally private and local, but access to transnational sources of improved strains, feed ingredients and vaccines, etc. has been important.

Milk

In the absence of cooling equipment milk was marketed only over short distances in the developing countries, by small scale private enterprises. Supplies remaining unsold or seasonally in surplus were processed into butter, ghee and sweatmeats. Where reliable supplies of fresh milk were not assured, consumers bought condensed milk in cans or milk powder from transnationals such as Nestlé. This led Nestlé into establishing processing plants where milk was obtainable at low cost from cows on natural grazing. Central milk plants to serve city populations were often the outcome of aid programmes initiated in the 1960s. Commonly they were municipal projects.

The basis of cooperative milk processing has been the prevailing urban orientation of pre-existing services. Rural milk producers needed an enterprise that put their interests first, i.e. assurance of a regular outlet for daily milkings. Recognition of the social significance of a regular cash income to small farmers brought in governments to establish and to support such systems.

Fish

Fish processing for local or more distant domestic markets e.g. drying in Indonesia, smoking on the West African coast and the shores of Lake Chad has been a private enterprise activity. Governments have set up parastatal and cooperative services in response to fishermen's complaints of exploitation, but in general they have not endured. In various parts of the developing world transnational enterprises have canned or frozen fish for export. They have also played a major part in the commercial production of fish and crustacea for processing. For this the warm waters and low labour costs of the tropical countries constitute an attractive environment.

SUPPLY OF RAW MATERIAL

The supply of raw material for agricultural and fish processing enterprises has three broad bases:

1. An effort to relieve fresh markets where prices are depressed by a glut of supplies.
2. A direct search for a low cost source of supply to serve an external market.
3. The need to keep a business running once an investment has been made to serve domestic consumers.

Gluts on the fresh market

While it is not generally recommended that a processing enterprise be established on the basis of periodic low prices for the raw material, many have this as their origin. There was a period in the 1950s when Israel faced occasional gluts of oranges on its markets for fresh fruit. It was advised not to make a major investment in processing as an outlet for distress supplies. Since then it has become a major exporter of citrus juices and other processed products. Its production base had the capacity to supply both fresh and processor outlets; growers benefited from the reduction of risk afforded by alternative outlets through the Citrus Marketing Board.

Dabaaga Canning Co. in Tanzania demonstrates how processing could be successful drawing on 'traditional' seasonal surpluses, when the enterprise was operated on an appropriate scale and with low overheads. From this base it was still possible to develop a domestic market and maintain effective promotion. Part of its supply came, however, from the operator's own farm; so its availability was assured, apart from normal production risks.

Kenya Cooperative Creameries had its origin in a need to find markets for milk surplus to local requirements. It was set up to process the surplus milk into products that could be sold in distant markets. After years of successful operation some of its plants became under utilized because of shifts in milk production relative to fluid market consumption.

Very low prices and evidence of overgrazing for lack of market incentive were major factors in government initiatives to set up livestock processing facilities in Botswana, Chad and Swaziland. These projects incurred periodic supply shortages due to disease and drought. There were, however, long periods of successful operation, notably in Botswana.

Many other plants have been set up in areas where the raw material was apparently in surplus to find that low yields set limits to production. Corn Products Corporation bought into Rifhan in the Pakistan Punjab because maize was available there and not used for human consumption. It offered reserve prices to clear the market; sufficient supplies were not forthcoming. It had to recruit qualified agronomic research and extension personnel, obtain new seed, and test out for suitability under local conditions potential improved production practices. It then had to finance growers to follow its recommendation (see Appendix 1 for the contract used).

Search for lower cost supplies

This has been a driving motivation for the transnationals. They faced rising costs of production and of labour in their home base. They looked to the developing countries for places where supplies for processing could be obtained more cheaply. For some crops they were obliged to organise the whole sequence of production. Del Monte in Kenya, Del Monte and Dolefil in the Philippines found that direct production of pineapples was preferable to purchasing from independent farmers even under contract. Uniform quality and state of maturation were critical for successful processing on a commercial scale. However, a proportion of pineapple supplied for canning in Thailand still comes from private farmers under contract. For many other items direct production of a proportion of the supply needed for efficient operation of the plant is considered desirable, to be assured of some supplies, to have supplies under direct control to fill timing gaps left by other suppliers, and as a basis for practical experience of production for use in transmitting know how to farmers under contract.

Processors using tree crops such as oranges and limes have the advantage that, once the trees have reached maturity, they can be expected to crop regularly for many years to come. Yields can decline, of course, if the price offered is too low to support pruning and other maintenance and can fall to zero if the price does not cover the cost of picking.

Diversion of supplies onto local fresh produce markets is a recurrent risk for the processor. For some crops such as grapes, limes and olives alternative outlets on fresh markets are limited in scope. This constitutes a protection for the processor. With many other products processing enterprises have found themselves in difficulties because the fresh market offered a much higher price. The possibility of expected price relationships being upset by external events is best foreseen in supply price agreements. Fixing a specific price in advance can impose strains on both the parties involved. An agreed relationship to a market indicator which reflects current conditions of production and of sale on alternative outlets, e.g. a major national market for fresh produce or an international commodity exchange, may be more convenient for both parties. Further alternatives include offering a basic price plus bonuses that vary with changes in the price of the processed product on designated markets. In any event the value of contracts with large numbers of suppliers depends on the good will of the parties involved. Legal enforcement is rarely practicable.

Much the same considerations bear on supply arrangements for vegetables. Conditions for specialized production must be favourable not only agro-climatically, but also in terms of human resources. Farmers must be willing to shift to intensive production to a degree often not previously envisaged. Tomato processing has been held back in some parts of Africa because offering a product that was within the price range of domestic consumers depended on farmers producing 20 tons of raw material to the hectare as against the five tons they considered a normal yield. Planning their supply base for tomato canning in Taiwan in the 1960s when it was new there, Kigoma looked for a group of farmers displaying willingness to adapt to new production patterns. Under their supply contract (summarized in the case study) the processor provided seedlings as well as other inputs on credit plus intensive technical assistance.

Full advantage should be taken of opportunities to extend the processing season and so reduce the incidence of plant overhead costs. Early, main crop and late maturing seedlings can be supplied. Supplies can be contracted from production zones with differing maturity dates according to soil, rainfall, location for sun and shade, drainage, access to irrigation, etc.

The production supply contract also enables the processor to specify varieties that are well adapted to his requirements, e.g. a high proportion of solid matter in tomatoes. These may differ significantly from those grown for fresh markets. Provision can also be made for specification of delivery times. This helps the processor keep his plant in continuing operation. It also reduces waiting time for producers' deliveries during the peak season and possible quality losses due to delays in acceptance at the plant.

Availability of low cost seasonal labour can be an

important consideration. This was critical in the picking of marigold heads in Ecuador. To obtain the maximum yield at the correct stage of maturity called for labour to go over the crop by hand several times during the season. The picking and drying of medicinal herbs is also a task for part time labour. Qualified assembling agents in production areas are needed to check the materials and ensure that they are properly dried. Chalam's Herbochen drew most of its supplies from a firm in Bombay which assembled materials from agents in different climatic zones. It was subject to sharp changes in supply prices due to other demands on the same wholesaler.

Sugar and tea are crops where the time period between harvesting and processing can only be short; otherwise there are substantial quality losses. This implies production near to the plant and in close integration with the processor. Because of the large supplies of sugar cane needed, its low value to bulk ratio, and the fact that it must be delivered to the plant within 48 hours of harvesting, cane to supply a factory is grown within a 20 to 30 km. distance. Traditionally this has been under the direct control of the processing enterprise. An additional consideration was the need to stagger planting and harvesting to distribute deliveries over as long a season as possible. On the other hand, the labour requirements of sugar cane harvesting were probably the heaviest made by any tropical crop grown commercially. A typical Cuban sugar mill in 1962 needed a minimum of 1,200 cutters to keep it fully operating. This is behind the long history of turbulent labour relations associated with sugar cane plantations. It has induced a shift to family scale growers convenient for access to the factory and linked to it by contract.

Competition between cassava processors in Thailand forced them to finance growers under contract and still pay the free market price. The cost of the credit had to be set against the saving on overhead cost through keeping the plant well supplied.

Need for raw materials to meet a domestic market opportunity

Rapidly growing urban populations and rising incomes have stimulated the provision of facilities to process produce for domestic consumption. With the investment made the processor must assure himself of supplies. Characteristically rice and maize millers locate themselves conveniently for access to producing areas. Growers may then bring their own produce to be milled for a fee. They can also offer it for sale to the miller on the spot. Where a larger throughput is needed to cover his overhead costs the miller must take direct initiatives to assure himself of continuing supplies. Commonly these include:

a) developing personal contacts with major producers;
b) establishing agents and sub agents to buy on his behalf on local assembly markets and directly from growers by visiting them in their villages. Such agents must be paid a commission, or allowed a margin on the price they have paid;
c) laying out credit to growers to help cover their cash outlays on production inputs, and possibly on family consumption and other obligations where social linkages provide an adequate guarantee of repayment via deliveries of grain for processing and sale.

Where pricing at the producer level is free, prices will tend to be low immediately after a harvest. The processor who can mobilize finance will take advantage of this to acquire stocks. These can then be drawn upon later to keep his mill running. The seasonal increase in their value will augment his operating margin.

If prices are stabilised by government intervention the independent processor can:

a) buy standard supplies at the official guaranteed price accepting the margin allowed for milling, with scope for extra earnings if his equipment gives higher than normal yields;
b) buy at higher prices varieties and qualities for which some consumers, and the distributors who serve them, are prepared to pay a premium;
c) buy at lower prices qualities that do not meet the government buyer's minimum standard, with the expectation that these supplies can be improved by drying, cleaning and blending.

Livestock feed mixing usually began as a profitable service to poultry raisers and milk producers prepared to pay for a product adapted to their needs and convenience. Basic materials for feed mixes have been coarse grains and milling residues unwanted for human consumption. Strategic for the mixer, however, was access to ingredients that would bring the feed up to a specific nutritional level. For this purpose he would buy fish or soy meal, or another high protein product from a specialised supplier or importer. To obtain some of the more advanced growth inducing ingredients he might need a link with a transnational.

Whether private indigenous, transnational or cooperative, processors serving domestic markets had generally to lay out money in advance to be sure of supplies. The private firms made their contacts with producers directly or through private agents. The cooperative unions used their local societies. Unilever Is used both the sunflower producers' cooperative in Thrace and independent oilseed crushers. In the dried fish trade in Indonesia credit went back to the original fishermen through an intricate channel of contributions and risk sharing. The state dairy enterprise in Bolivia made credit available in kind. Straits Fish Meal paid only a market clearing price for their raw material, but to be sure of supplies it financed many fishermen.

MARKETING THE PRODUCT

Export markets

Here the transnational/joint venture usually has the advantage of a base in the market envisaged. It is aware of entry requirements and consumer preferences for product quality attributes, packaging and presentation. It may have an established brand and its own distribution network. Orange juice is exported in bulk from Brazil to be repacked in Germany by the same enterprise that owns the plant, likewise lime juice from Ghana destined for Cadbury Schweppes in the UK. Japanese partners ensured the smooth entry into their market of tomato products from Taiwan, frozen chicken parts and sugar from Thailand.

A specific comparison of the export prices achieved by transnational and independent pineapple canneries in the same country (Jabbar S. 1972 'A case study of production orientation: Malayan canned pineapple.' European Journal of Marketing (6) 3) showed the operation linked to a transnational to have a distinct advantage. The independent plants obtained much the same price as it did when the market was good. But in face of a down turn they could sell only at a discount.

New independent enterprises with a product to export begin by using brokers. It is their business to know which distributing or retail chain will see an advantage in buying a particular product. If the product is attractive they can enable the processor to skim the market obtaining the highest prices. Basotho Canners were able to do this with purple headed asparagus much liked by German consumers.

As the quantities offered increase and competition intensifies processors seek firm contracts for at least a part of their output. The smaller independent orange juice processors in Brazil and fish meal processors in Peru eventually grouped together for joint sales arrangements. This occurred once it became clear that no one of them had much chance of establishing a dominant position.

Regular sales of high value on an export market may warrant setting up a branch there to keep in continuing contact with customers. Botswana Meat Commission did this in the UK and the Société de Commercialisation du Poisson in Las Palmas and Japan. State trading companies handling a range of products can also do this economically. It was Internor, the North American subsidiary of Interbras that organised to advantage the export of lobster tails from Brazil. By 1986 it was estimated, (Kirchback, F. 1986, Structural change and export marketing channels in developing countries, Geneva, I.T.C.) that national enterprises including state trading organisations such as Interbras, had a larger share of total exports from developing countries than the transnationals.

Domestic markets

The traditional channel to domestic markets is through established wholesalers with their own links to retail outlets. This relieves the processor of the burden of looking for more direct outlets, and perhaps more important, of collecting payments as they become due. Enebor used the market at Illushi which attracted wholesalers from other areas. It was a coup both for Haji Mansur in Indonesia and Hanapi and Sons, Malaysia when they found a big buyer. With his contract in hand they had no difficulty in obtaining finance from the bank. The parastatal and cooperative processors generally followed the same course. Under the Anand system in India the district union left to the cooperative federation the responsibility of finding markets for its products.

A combination of branch outlets and retail sales agencies is used by processors who want to follow their products in distribution. Ratna Feed Industries supplied regular dealers with stocks on credit allowing them a commission of five percent. It also used them to assemble eggs produced by its feed customers. Unilever Is developed a similar pattern for its edible oil products in Turkey. Dominant considerations were to control closely both the price charged to customers and the quality of the product on offer.

Direct sales to supermarkets and specialized restaurants were preferred by the integrated broiler processors in Jamaica and Thailand. These buyers took substantial quantities and had the freezer cabinets to maintain the necessary low holding temperature. Small scale processors may both wholesale and retail their products like the first broiler raisers in Lebanon.

Large scale retailers setting up a processing operation tend to find that its product flow matches only to a degree that needed by its retail customers. Thus La Favorita supermarket chain acquired its own abattoir with the intention of paying a premium to livestock raisers to supply high quality animals. It found that it had to work with private wholesalers who had outlets for the parts of the animals its retail shops did not want.

CONCLUSIONS

The tendency for particular forms of enterprise to be found more frequently in the marketing of certain products also holds good for processing associated with marketing. This does not mean that other kinds of enterprises have no prospect of success. It does suggest that their planning and management will need special attention.

Private enterprises are quick to seize new profit opportunities. They have a direct incentive to keep down costs. Often they have valuable local knowledge, experience and contacts. They can constitute both new initiative and new competition. However, in the absence of competition they may pursue their own interest to a degree that is resented by trading partners in a weaker bargaining position. There is a role here for mediation by a qualified government department.

Knowledge of what must be done to meet export market

requirements and of the technologies required for effective processing is now widely available. National enterprises are gaining on the transnationals in the export marketing of agriculturally based products. Nevertheless, the assured outlet of a repacking plant, distribution channel, or established brand under the same management is a continuing advantage. Transnationals can also contribute highly productive ingredients for use in processing for domestic markets; this can be on a contract basis.

An effort to organise their own processing cooperatively has often been the protest reaction of farmers who felt exploited by existing market channels. Under developing country conditions it seems to have best chances of success where the processes are standard and sale to a pre-established outlet is assured.

Parastatal processors are often established to implement an official policy that requires continuing government support. This can be to help stabilise prices, or where access to favourable markets depends on government negotiations and guarantees. The allocation by government of monopoly powers, external aid resources and technical assistance has also been instrumental in the success of processing projects designed to help very small farmers. Important considerations are that the price policies pursued be realistic and that the enterprise be accountable to its users. Indefinite continuance of parastatal and cooperative monopolies can impede, however, beneficial adaptation to changing demand structures and process technology.

It is sometimes said that private enterprise is quick to take up new lines of processing such as integrated broiler operations, and that new initiatives in processing the output of traditional agriculture await government support. The issue is seen more clearly in feasibility terms. Where private initiative in modernising rice mills is lacking, it is often because official pricing does not provide an incentive. The effective marketing of meat from range cattle in Africa may depend on the negotiation of market entry and maintenance of extensive disease controls, obligations that only a government can shoulder. It must also be recognized that the establishment of new integrated production, processing and marketing requires access to substantial capital. Private enterprise can only go ahead if funds are available from local financing institutions or conditions favour their entry from elsewhere.

Five elements seem to be needed for the successful operation of any processing/marketing system:

1. Reduction of risk in terms of both price and volume, through formal or informal arrangements that ensure both an outlet to farmers and fishermen and continuing supplies of products to processors and distributors.
2. Mutual dependence and interest to ensure continuity and risk sharing among the parties entering agreements either contractual or informal. The more perishable the product the more need there is to reduce risk through complementarity between producer, processor and other handlers.
3. An efficient system of intelligence to identify market opportunities and technology options and applications.
4. Effective transfer of knowledge via close working contacts between those possessing the technology and the users, producers or processors.
5. Professionally competent management with an imaginative understanding of the various parties with which it deals.

ISSUES FOR DISCUSSION

1. Which areas of agricultural and fish processing are occupied by indigenous private enterprises in your country? How do you rate their performance? What are the handicaps they face in

expanding their operations?

2. Prepare profiles of some transnational/joint venture processing projects in your country. How were they initiated? What advantages have they over possible competitors? How do you rate their performance?

3. What agricultural and fish processing is undertaken by cooperatives in your country? Have they developed in response to specific disadvantages felt by primary producers? What support and protection do they receive from the government? How do you rate their performance?

4. What is the role of parastatal enterprises in the processing of farm and fishery products in your country? In response to what considerations were these enterprises established? How do you rate their performance?

5. Comment on the review of the performance of alternative kinds of processing enterprises presented in this chapter in relation to: a) commodities handled; b) assembly of raw material for processing; c) marketing of the processed product. How far is it supported by experience in your country?

6. Do the conclusions presented at the end of this chapter have general validity for your country? How would you modify them to reflect better your conditions?

FURTHER READING

Abbott, J.C., (1987) Agricultural marketing enterprises for the developing world, Cambridge, Cambridge University Press.

Arhin, K. et al., (1984) Marketing boards in tropical Africa, London, Kegan Paul.

COPAC (1984) Commodity marketing through cooperatives: some experiences from Africa and Asia and some lessons for the future, Rome, FAO.
George, S., (1977) How the other half dies: the real reasons for world hunger, Montclair, New York, Allanheld, Osmun & Co.
Harper M. and R. Kavura, (1982) The private entrepreneur and rural development, Rome, FAO.
Jones, W.O., (1972) Marketing staple food crops in tropical Africa, Ithaca, Cornell University Press.
Lall, S., (1984) the multinational corporation, London, Macmillan.
Lall, S. and P. Streeten, (1977) Foreign investment, transnationals and developing countries, London, Macmillan.
Lappe, F.M. and J. Collins, (1978) Food first: beyond the myth of scarcity, Boston, Houghton Miflin.
Moran, T.H., Ed. (1986) Investing in development: new roles for private capital, Washington, Overseas Development Institute.
UNCTC (1981) Transnational corporations in food and beverage processing, New York, UN.
Schluter, M., (1984) Constraints on Kenya's food and beverage exports, Nairobi, Institute for Development Studies, University of Nairobi.
Van der Laan, H.L., (1975) The Lebanese traders in Sierra Leone, The Hague, Mouton.
Widstrand, C.G., Ed. (1970) Cooperatives and rural development in East Africa, New York, Africana Publishing House.

4 Processors' contributions to development and the role of government

The theme of this chapter is that agricultural and fisheries processing enterprises are potent motors for development. Their contributions will be the more easily made and the more far reaching where they receive clear support from governments. These contributions include:

a) mobilizing finance, inputs and technical assistance for agriculture and fishing;
b) inculcating new more efficient production techniques;
c) raising farmers' and fishermen's incomes and promoting further investment in agriculture and fishing;
d) creating employment;
e) transmitting technology and management capacity;
f) reducing the time spent on food preparation in the home;
g) reducing waste and extending the period of availability of traditional foods and broadening the range of foods offered to consumers;

h) earning valuable foreign exchange;
i) contributing to local and central government revenues through taxes.

These contributions will accrue more quickly and be more far reaching if processing enterprises are backed up by national governments. Positive government policies towards such enterprises provide for:

1. financial, fiscal and other facilities for their establishment;
2. assistance in developing physical and social infrastructure needed for successful operation;
3. ongoing public services that complement the activities of the processor;
4. foreign exchange access, import licenses and exemptions from import duties as needed to obtain supplies, spare parts, etc., essential for the processing operations.

These contributions processors can make to development and the kinds of public support needed will be examined further in the following chapter. We shall then take up welfare aspects of agricultural processor activities and the measures that can be taken by governments to promote a more equitable distribution of benefits from processing and to protect environmental and other long run public concerns.

COORDINATION OF FINANCE, INPUTS, TECHNICAL ASSISTANCE AND CASH INCENTIVES FOR FARMING AND FISHING

Processors have taken a lead in coordinating these services to producers both to develop a market for their own products and to be sure of suitable supplies. The first of these roles has frequently been assumed by the feed processing industry. Ratna Feed was a pioneer in Nepal. In the 1970s it was instrumental in the growth of many egg and meat poultry farms.

Intensive broiler production has developed since through integration with the feed supplier. Maize millers in Ecuador had a promotional role in some of the shrimp production enterprises. To be sure of obtaining a suitable flow of raw material, processors have channelled productive resources to their suppliers in appropriate proportions. They consolidated efficient use of these inputs by providing an immediate market outlet. This is widely evidenced in the case studies presented in Chapter 2.

Producers associated with processors usually benefit directly or indirectly from their market intelligence. This reaches processors through their foreign clients as in the case of Allana Bombay, through brokers or through the sales agencies of state trading organisations - Internor for lobster tails in Brazil. The strict quality standards and controls applied by processors are a guide to producers, likewise adaptations in product size and other attributes to meet specific entry conditions and consumer preferences of export markets.

The producer processor supply contract has been found an effective mechanism for ensuring, on the one hand that supplies meet processors' requirement and in return, guaranteeing the producer a convenient outlet. The advantages have been clear. Such contracts have provided:

1. an assured market for producers' output;
2. access to the company's technological and other services; this involved field advice and training, supply of seeds, fertilizers, and other inputs as needed to secure production at low cost with high quality standards; and
3. easier access to credit; loans have often been made to farmers at less than bank rates of interest and for longer periods; in cases where the firm did not provide the loans, banks generally accepted a contract as collateral.

In contrast, the common system of supplying credit for inputs through a financial intermediary, extension through a government agency and buying by a marketing board has encountered problems of input timing, coordination, payment and collection. The contrast is least evident with grain crops grown by large numbers of farmers and offering more economy to generalized extension and input distribution.

On the other hand, having entered into a contract for a specialized product farmers may be unable to exercise further choice. The captive nature of such contracts depends on their prices for inputs and output relative to those of competing enterprises. The newer a product to the country, the more limited are the alternative options. Another problem can be the manipulation of quality standards where contracts specify that these are the sole domain of the buyer. A firm can raise its quality standards not only to reduce the quantity of raw material accepted but also to obtain a portion of it at a very low price. The situation may also arise when the firm does not comply with its commitment to purchase all supplies contracted because it foresaw losses on their sale when processed. In the worst case, the firm may close down leaving the producer without alternative markets or the means to continue operations.

Indications that a major processing enterprise is facing financial difficulties constitute a cue for continuing surveillance by a qualified government department. Action alternatives range through:

1. specific assistance to help it overcome a short term crisis;
2. assuring some longer run improvement in its operational environment;
3. preparing alternative outlets for production resources should eventual failure of the enterprise seem probable.

Common short run problems of processing enterprises are inability to obtain essential supplies, disputes with raw material producers over prices and quality definition, disputes with employees over wages and conditions of work, and interruptions in access to established market outlets. Most of these are amenable to a positive government intervention. It can expedite allocation of supplies. It can mediate in disputes with producers and labour. It can help find solutions to transport breakdowns. It can negotiate with other governments to re-open access to markets, etc.

An important potential role of government is in arbitrating over prices to producers supplying processors under contracts. Specialized marketing boards have been used for this in some countries. In Canada, for example, boards were established specifically to represent the producers of crops for canning vis à vis the canners. Annually, they would negotiate with the canners the terms on which produce would be supplied, prices, quantities, quality specifications, timing, etc. They were empowered to allocate quotas to individual growers. After discussions with producers and processors a government department has determined equitable prices for asparagus and mushrooms in Taiwan. In Brazil, the government has required the negotiation annually of a single producer to processor price for juice oranges.

Such an intervention must take into account the economic realities of processing. An indicative price can be useful, provided it is regarded as an average price. The processor should be able to reduce it in peak harvesting season and raise it progressively before and after the peak. In this way he provides an incentive to producers to extend the supply season. Similarly, there should be scope for offering a premium for produce fitting closely his specifications, or higher in processing yield, and lower prices for less valuable deliveries.

Rigid government controls can quickly work against efficiency in processing. In Tunisia, in the mid 1980s, tomato paste plants averaged only 35 days of operation annually, as against an international norm of around 100 days. Inflexible raw material pricing provided no incentive to producers to extend their delivery season. The sales price set for the processed product was based on operating costs plus a normal margin, so additional plants were installed to meet the peak seasonal load. The result was low capacity utilization and high costs for all the tomato processing plants, likewise for those canning pepper paste and fish. The processed products were expensive for domestic consumers and uncompetitive on export markets.

Rising production costs in the early 1970s obliged sugar cane processors in Kenya to pay higher prices for their raw material. However, the government kept down the price of sugar to restrain inflationary pressures because it was an item in general consumption. The government persisted in this until the sugar processors of the country were virtually bankrupt. The government received no dividends from its own shareholdings in sugar processing, but its income from excise duties increased since consumption was favoured by the low price.

At one time the vegetable cooking oil and margarine of Unilever Is in Turkey seemed likely to be classified as manufactured products rather than food. This meant that they would be subject to a high sales tax. Only intensive lobbying with arguments founded on accepted procedures in other countries secured a reversal of this classification.

Endemic problems in the supply of raw materials for processing due to pests and diseases, lack of or high cost of essential production inputs, and official policies to promote the output of products competing for the same resources, are best approached through open negotiations between the processor and the responsible government department. Solutions are found most easily where there is full understanding of the impact of costs and other constraints on the processor's operation on the part of the government, and of producer, consumer and national economic interests on the part of the processor.

The disappearance or sharp reduction of the market open to a processor necessarily calls for processor/government collaboration in relieving the situation.

Thus in 1987 the quantity of Tunisian olive oil that would be accepted by the EEC countries, its main market, was fixed at 46,000 tons. This was far below Tunisia's export potential. The course of action followed was:

1. Provision of credits to regenerate existing plantings in order to raise yields so that growers could accept lower unit prices.
2. Provision of credits to install new olive presses so that the average quality of oil delivered under the quota would be higher and so bring a higher return.
3. Efforts to develop new markets in North America, Eastern Europe and the Middle East would be assisted by subsidization of transport and promotional costs during the initial years.

The decision of Gulf and Western Inc. to withdraw from its sugar operations in the Dominican Republic in 1984 was regarded as a characteristically selfish action on the part of a transnational. Already for some years, low sugar prices and high taxes had resulted in a low return on its investment there. The decision to withdraw was taken only after the death of a chief executive who had a strong sense of obligation to the Dominican Republic. In the event another company took over the Gulf and Western holdings; one of its first actions was to diversify into pineapple for the fresh market.

TRANSMISSION OF TECHNOLOGY AND MANAGERIAL CAPACITY

The most direct vehicle for the transmission of technology by processors has been the production/marketing contract. Farmers with no prior experience of a crop have been supplied with a proven package of inputs and husbandry procedures backed up by credit and a guaranteed market outlet, as for asparagus in Lesotho and marigolds in Ecuador, or protection against risk as in broiler contracts in Thailand and Jamaica. They have adopted successfully the new technology. The Siam Food Products case has been included as a warning. It shows the difficulties that can arise when the technological package is not well adapted to local conditions or work methods.

Standard technology can also be acquired quickly enough without such contracts if the price incentive is sufficiently attractive. This was the experience of the Lebanon broiler boom, the frozen orange juice development in Brazil and fishmeal in Peru. Seeing their neighbours become rich was stimulus enough for others to learn and emulate.

A producer cooperative processing enterprise can be an effective frame for the transmission of technology to members and staff through group pressures and individual participation in policy and decision making. This was a feature of the Anand cooperative model in India. Chitali shows that a family enterprise can also be an effective vehicle for technology transmission to large numbers of suppliers, at less economic cost than the cooperatives taking into account their tax exemptions and subsidies.

Government support seems to be especially important when:

a) the investment in the productive resource for processing is very long term; government credit financed the lime and citrus plantations in Ghana and Brazil;
b) the support services needed call for authoritative backing and maintenance on a country-wide basis over long periods, like the foot and mouth disease control measures in Botswana.

The support of government advisory, veterinary and disease control services can also contribute greatly to the extension of new technology more generally, as in the Lebanon broiler boom where vaccines distributed free of charge reduced greatly the risk of epidemic disease. Often, however, a specific additional input by a processor may be required if full advantage is to be taken of existing institutional services. The research and development undertaken by governments and their extension service messages tend to be too general. They are insufficiently adapted to the needs of particular sets of producers and to the conditions under which they operate. Offering a financial incentive to farming and fishing enterprises to adopt a new technology through a contract can be decisive in their breaking out of a static situation.

The State Government of West Bangal has promoted modernization of rice mills through provision of subsidized working capital and a direct subsidy of 15 percent of the cost of new machinery. While communist in its political orientation it supported innovation by large private mills. Monopoly wholesale rights for defined areas were assigned to them to ensure capacity utilization after modernization. The State saw social benefits from the creation of a concentrated, monitorable rice processing and distribution system.

In general, processors have found it much easier to introduce new technologies as a complete package associated with a new product than to improve the handling and management of traditional agricultural products. In this area government intervention may be needed to promote new processing technology. Frozen broilers are the classic example of the new production/processing/marketing system introduced as a whole. In contrast, the marketing potential of traditional livestock in Botswana, Chad and Swaziland was only exploited when government secured financing for the new abattoirs required and backed it up with disease control and other support measures.

However, there are many variations on this theme. The Mazenod asparagus project in Lesotho was a government initiative to take advantage of a protected market in the EEC that it had negotiated. Some of the most striking innovations such as the sale in the Gulf States of frozen fish and meat from India were private enterprise initiatives drawing on traditional sources of supply. Recognition of a market opportunity and application of the appropriate technology to meet the export market standards were the important factors here.

Physical inputs such as facilities, basic stock, boats, nets and fuel, refrigeration for production/ processing of fish and crustacea are rather costly and capital intensive in most developing countries. Many of these inputs have been provided directly or financed by processors. The joint venture between local and foreign interests has been an efficient way of acquiring physical technology and of inducing technological advances. Striking examples include the provision of new facilities for ice making, processing and freezing, increasing offshore catches at Nouhadibou in Mauritania, the development of shrimp cultivation and processing in Ecuador and the acquisition of specialized techniques for drying jelly fish in Indonesia. In contrast are the problems encountered for lack of foreign exchange and access to market technology in Senegal. (UNCTAD 1984 Industrie alimentaire au Senegal, en particulier la transformation des céreales, produits de pèche et produits laitiers, Geneva)

With foreign exchange constraints a continuing brake on private ventures, many developing countries look to foreign aid as a source of capital for new processing installations. Concern to make use of aid budgets, or of bilateral offers in kind has resulted in some of the more conspicuous white elephants. Fruit, vegetable and meat canning plants have been built that have proved much too large for the volume of produce available within economical reach. Centralized management, where this is undertaken by government, and inflexible marketing policies can lead to high collection and distribution costs, as evidenced in the Bolivia Dairy Industries Enterprise case. Government support of smaller locally based initiatives is often preferable. They are better placed to adapt to changing supply and marketing conditions.

The case studies show that many different mechanisms have been used for the diffusion of technology. Irrespective of the mechanism, the technology package has to be adapted to the socio economic environment in which it is to operate. Its diffusion evolves best through interactions and reaction with progressive minded individuals and leaders among farmers and fishermen. Emulation will spread following demonstrated success. For this reason, one of the surest methods of transmitting knowledge and experience is through close working contacts with enterprises or technicians possessing suitable advanced technology. The process is hastened and consolidated where there is a direct shared financial incentive. This stands out in joint venture fish catching in Mauritania and in the Arbor Acres partnership with the Charoen Pokphan Group in Thailand.

Employment for a time of qualified expatriate managers and technicians by a national processing enterprise has also proved effective, e.g. jelly fish drying in Indonesia, livestock and meat processing in Botswana and Chad.

During its first years Basotho Canners received technical and managerial assistance under the FAO/UNDP aid programme. This can also be provided under managerial contracts. These may have some advantages over technical assistance:

1. The manager has greater authority and responsibility in dealing with all matters conducive or prejudical to performance.
2. Payment can include incentives for performance as well as basic fees; this secures a commitment close to that of an owner.
3. In the case of government sponsored projects it may pave the way for equity participation by enabling a potential partner to gain first hand knowledge of the enterprise and its environment. Delegation of management responsibility also implies a willingness to grant a level of autonomy that will be beneficial to subsequent operation.

The contract made by the Government of Kenya with Booker McConnell for the management of the Mumias Sugar Co. is often quoted as a model of its kind. (Abbott, 1987)

Familiarity with advanced technology and marketing initiative has also been brought into developing countries by individuals who found conditions favourable to their becoming long term residents or citizens. These include the founders of the Dabaaga Canning Co. in Tanzania, of the marigold venture in Ecuador and of the rapidly developing export of frozen shell fish from Thailand. When local technological capacity is fairly high, highly productive stock, feed ingredients, inoculants, etc., incorporating accumulated technology can be acquired from an external source and multiplied in the developing country, as in the broiler cases in Jamaica, Thailand, Lebanon. Many such technologies can then be transferred to the developing country under licensing and purchase arrangements.

A continuing responsibility of government in technology and management transfer is to institutionalize it. Business management can be taught at universities and at intermediate levels. Marketing components in management and agricultural economic teaching programmes can be strengthened. Food and agricultural product technology research and training can be featured at universities and technical colleges. Degree courses incorporating these three subjects can be an option for students seeking a processing enterprise specialization. Short seminars focusing on current issues can be designed for processing practitioners. Here, and in courses for students, practical talks by people working locally in processing and marketing should have a priority. Too often teaching courses are set up in developing countries, but remain theoretical, or are based on experience of quite different conditions. Links with ongoing commercial enterprises can be strengthened further by inclusion of their representatives on committees to discuss teaching course content. They can also be invited to support scholarships, finance research oriented to their needs, and endow professorships.

FOREIGN EXCHANGE EARNINGS

Striking examples of foreign exchange earnings from exports by agricultural and fish processing and marketing enterprises are:

> $1,400 million frozen orange juice concentrate, Brazil 1984;
> $600 million cassava pellets, Thailand 1983;
> $148 million frozen shrimps, Ecuador 1982;
> $148 million frozen fish, Mauritania 1983;
> $70 million fishmeal, Peru 1964;
> $43 million frozen chicken, Thailand 1981.

Some of these figures reflect boom conditions and have not been maintained. There are many other cases, however, where foreign exchange earnings from a processed product have continued at a substantial level over the years. Illustrative are meat for Botswana, lime juice for Ghana, tea for Kenya. For some countries with very limited export prospects, earnings from processed agricultural products are strategically important for their economies. The cases presented in Chapter 2 only illustrate the position. For a wide range of developing countries exports of agricultural and fishery products after processing constitute a very large part of their foreign exchange earnings, e.g. Ethiopia coffee, hides and skins; Uganda coffee and cotton; Malawi coffee, tea and tobacco.

Achieving these exports may involve some prior imports, as of processing equipment, containers or material for the local manufacture of containers, specialized supplies for processing. Some inputs needed for the production of suitable raw materials may also have to be imported. Rarely, however, with agricultural and fisheries based exports do the associated import requirements cancel out much of the foreign exchange gain.

Transnational processors tend to come with in built export outlets. They constitute, however, only a part of the potential processed agricultural product export base. The share of national enterprises is now larger. They can benefit greatly from an intensive government export promotion programme.

Export promotion

Government assistance in an export drive can take three main forms:

1. orientation and supplementation of existing official institutions to maintain effective support;
2. provision of financial incentives and assistance to independent export promotion;
3. elimination of obstacles to exporting.

These approaches may be illustrated from the export promotion drive initiated in Tunisia in the mid 1980s. Olive oil, canned fish and fruit and vegetables and wine were among the products involved.

<u>Export promotion centre</u>. Such a body had been established much earlier, as in many other developing countries. Its primary activities were to:

a) study export markets and provide information on them to potential exporters;
b) establish contacts between exporters and potential importers;
c) promote Tunisian products at trade shows and fairs; and
d) advise the government on trade policies, simplification of export procedures, etc.

In practice this body worked mainly with the commercial secretaries at Tunisian embassies. It was not rated very effective. In 1985 it was assigned an export promotion fund derived from a levy on certain imports. This fund could be used to:

a) subsidize the transport costs of sales to new markets, or of new products to traditional markets;
b) cover half the cost of specific research for sales on new markets.

Export trading companies. Handling a range of export products such enterprises can maintain a continuous presence on export markets. This may not be feasible for a processing firm with only one product or with limited quantities to offer. To benefit from competitive efforts the government raised the number of such firms with special export and foreign exchange facilities to seven.

Export financing. Up to 80 percent of a firm's export order can be discounted at a commercial bank at six percent interest. In turn the commercial bank can rediscount its credit at the central bank at four percent interest. This helps enterprises finance the production, processing and dispatch of materials for export.

Export credit insurance. A new company with commercial bank and insurance company participation was set up for this purpose. It insured the exporter against non payment for his products for commercial, political and catastrophic reasons, at rates of 0.5 to 0.9 percent according to the risk foreseen.

Federation of exporters. Under the auspices of the Chamber of Commerce this body provided:

a) information and training to private exporters;
b) a concerted voice to the government on policies, laws and procedures.

Elimination of obstacles. This included:

a) allocation of foreign exchange for imports of processing supplies, packaging materials, etc. needed for sales on competitive external markets;
b) rebate of import duties on such materials when re-exported;
c) limitation of export controls to observance of minimum quality standards and determination of value for foreign currency repatriation controls where these are judged necessary.

Prompt action on requests for export credits, exemptions from duties, etc. is essential so that contracts can be fulfilled and incoming supplies are on hand when needed. Uncertainty over the arrival of essential supplies often obliged processors in Tunisia to carry extra stocks, so adding to their costs.

TAXES AND PUBLIC SERVICES

The contributions to public revenues made by processing enterprises through taxes are concrete and can be very substantial. The rate on its profits foreseen by Jamhuri Tannery in Tanzania was 40 percent. Unilever Is paid $7 million in tax to the Government of Turkey in 1984. Taxes paid by the Meat Commission in Botswana ($9 million, 1982) were one of the government's main sources of revenue for many years.

To promote the establishment of new processing enterprises most governments recognize the need to ease their tax burden during the initial years. To attract a major new investment, governments may agree to forego direct taxes for a specific period, say five years. They may also free the enterprise of import duties on equipment that must be imported. Acceptance as costs deductible from profits before taxes of generous depreciation rates for facilities and equipment is another concession that is important for capital intensive projects.

In most developing countries new processing enterprises are eligible for medium term loans from development banks at concessional rates of interest. They may also be able to obtain short term financing on favourable terms for defined periods. Such finance should not be based simply on the collateral offered. Financing agencies should be equipped to help processors formulate sound projects and to monitor their use of borrowed funds.

Cooperative processing enterprises often benefit from legislation that enables them to distribute surpluses without payment of income tax.

Collaboration from central and local government will also be needed in the provision of essential infrastructure. Location of a processing enterprise conveniently for production of the raw material it needs may take it to a distant thinly populated part of a country. Transport facilities, water and electricity supplies, housing, schools, hospitals, policing may be lacking. In the past wealthy transnationals have been able to finance all these, like Dolefil in Mindanao. More often this is not feasible. If the enterprise is to operate, and the living conditions of its employees are to be acceptable, central and local government must carry a large part of the cost. They will be compensated later, directly via taxes, and indirectly through increased economic activity in the area.

The rationale for government assistance in the establishment of processing enterprises is illustrated in Table 4.1. Hypothetical figures are used in comparing commercial or financial and overall economic annual returns on a project. The firm receives payment for its exports at the official rate of exchange. The foreign exchange goes to the central bank. In the economic appraisal these earnings are increased by a factor of 30 percent to reflect its scarcity. Additional earnings of suppliers of raw materials and services to the plant estimated at $50,000 are

Table 4.1 Illustrative comparison of financial and economic returns

	Financial criteria	Economic criteria
	(. . . . dollars	. . . .)
<u>Gross income</u>		
Export sales	500,000	650,000
Domestic sales	500,000	500,000
Additional earnings of raw material suppliers	-	50,000
Total income	1,000,000	1,200,000
<u>Costs</u>		
Labour	300,000	250,000
Taxes	50,000	10,000
Rent of site	-	20,000
Raw material, other supplies depreciation, maintenance, etc.	550,000	550,000
Total costs	900,000	830,000
Net income	100,000	370,000

income to the local economy. The cost of workers employed by the plant is reduced to reflect what they would be likely to earn if there were no plant. Of the taxes paid it is estimated that only $10,000 are spent on services to the plant; the balance is a net gain to the government. For this project the government had a site which it made available free. Its economic account shows a debit of $20,000 that might have been obtained for the site if it were rented for an alternative use.

If the capital invested were $1,000,000 the return

of 10 percent would be barely acceptable on a commercial basis. On the economic account, however, the return is over 30 percent justifying government provision of a site free. The project rates high because it purports to earn scarce foreign exchange, put to work under used labour and be a valuable tax payer. Under different circumstances the position might be reversed. The commercial profitability criterion might over state the economic benefits from a national view point with correspondingly lower priority for government support.

Inspection of processing plants and of the products they offer for sale for conformity to public health requirements is a normal government responsibility. This requires the recruitment of qualified personnel and access to a laboratory for the testing of samples. In the case of processing for export the government may also inspect for maintenance of minimum quality standards. This service is intended to ensure that products sold by a particular processor do not give a bad name to those of the country as a whole.

The pattern of support and control services that can be provided by a government concerned to promote an effective processing industry is well illustrated by that of Thailand:

1. A government sponsored investment board puts out information on potential investment opportunities and seeks out likely investors and partners.
2. The Ministry of Agriculture stands ready to advise on the availability of suitable raw materials and to assist in their development and provision.
3. The Ministry of Commerce and Industry specifies technical standards for products affecting processing such as metal containers; it inspects plants for quality of processed product, safety considerations and waste treatment systems.
4. The Ministry of Public Health implements laws and regulations to protect consumers:

it monitors the quality of products offered on the market.

5. The Ministry of Education is responsible for education and training in food science and technology, business management, marketing and other relevant subjects.

Specific assistance to new agricultural and fish processing enterprises includes:

a) low cost long term development finance;
b) short term finance for exports at seven percent interest;
c) exemption from import duties of equipment and supplies needed in processing local raw materials for export;
d) rebates of taxes on processed food exports;
e) reduced charges for energy used in processing materials for export.

It is also important that the application of licensing and inspection requirements make allowance for small enterprises serving local markets. Requirements of legislation designed for sophisticated industry can be a hurdle difficult for them to surmount. In many countries a new processing initiative must observe:

a) requirements to obtain various permits and licenses;
b) labour laws;
c) factory safety regulations;
d) public health protection controls;
e) income and excise tax laws.

To ease the way for the smaller processors it is usual to apply these controls in full only when the paid employees exceed a particular number e.g. 15 or 25.

SOCIO-ECONOMIC ASPECTS

By conserving perishable produce that might otherwise be wasted, processing adds to the productive output of a country and to the income of the producers concerned. It has been shown that enterprises processing produce for export markets can be potent earners of foreign exchange. Wisely spent, these earnings benefit the domestic community as a whole through provision of development equipment, fuel and other essentials, and incentive consumer goods.

Agro industries are major providers of employment. This is particularly important for the lowest income countries. The average rate of growth of such employment in these countries 1970-75 was 7.9 percent, far exceeding the rate of population increase 2.8 percent. Agro industries also provide important employment opportunities for women. In India 60 percent of the workers in the tobacco industry are women.

Processing for domestic markets tends to serve medium and high income consumer groups able to afford prices that cover the cost of processing. Often this substitutes for imports, so freeing foreign exchange for other uses. Some processed products, however, dried and smoked fish, salted sardines, canned mackerel and tomato paste improve directly the diet of lower income groups. For many millions of women in Asia and Africa the milling of rice and maize and the commercial processing of cassava relieves them of the drudgery of hand pounding and similar home treatments.

Contract growing of a specific crop for processing has been criticized for leaving farmers' families without land for food crops. Local marketing channels often lag behind in attracting food supplies from other areas. Most contract cultivation programmes now foresee this need. Food crops are incorporated into the rotation. Otherwise, land is set aside for subsistence cropping. Generally, food crop production improves where there is cropping for processors because farmers learn the advantages of applying fertilizers and other inputs. A processor supplying farmers under contract with fertilizers for use on a specific crop can also add some more fertilizer for use on food crops. The cost can be set against payment for the crop contracted for processing, provided this is of sufficient value. The Kenya Tea Development Authority was considering such an arrangement in the early 1980s.

Processors have also been criticized for favouring larger farmers as suppliers of their raw material. In fact, for some crops smallholders offer advantages through the direct incentive for care in harvesting. This was the case with tea in Kenya and marigolds in Ecuador, also tobacco in various countries. Reduced risk of stoppages due to organized labour has led many sugar processors to favour production by smallholders. With some other crops contracting with a large number of small farmers necessarily raises the extension and management costs of the processor without compensating benefits. Where this is desirable on social grounds, governments can provide for it in the set of conditions and inducements governing establishment of a project. When it set up a vegetable freezing plant at Cisterna in central Italy, Findus agreed to contract for half of its supplies from 100 small farmers; the balance was to be produced by one big farm. It did this for good public relations.

Production of the raw material for a processing enterprise is sometimes concentrated in a relatively small area. The sharply rising incomes of the direct participants is then said to raise the cost of living for the rest of the population in the area. Dolefil has been criticized for this. (Tavis, 1982) The immediate beneficiaries of an enclave type processing operation may indeed be 200 to 400 families, plus the direct and indirect employees of the processor. In fact, the benefits of intensive pineapple and sugar production, broiler raising, shrimp farming, etc., extend much further in that they support new demands for housing materials, consumer goods, feed grain and a range of services. Satisfying market demands resulting from increased income in these related sectors extends further the indirect benefits.

Some processing systems such as cooperative dairies have the specific advantage of widening participation in marketing activities. The Anand model in India trains cooperative workers and farmers on the job. Through its activities, quality control and value analysis of milk and its components are becoming increasingly village level technologies. Particular stress is laid on organising farmers to set up and run their own societies and to upgrade dairy animal quality and related agricultural production. Cattle owners in thousands benefit from milk and meat processing initiatives where these expand the market for traditional suppliers as in Bolivia, Botswana, Chad and Kenya. The productivity of traditional herds has increased steadily in Botswana in response to the sure market offered by the Meat Commission.

Even in such cases, however, there are losers as well as gainers. The mobilization of feed resources for small scale commercial milk production in India, for example, absorbs grass and crop residues formerly available free to lower segments of the rural population. This may be the immediate result; in the short run the advance of one sector of a rural economy may be to the disadvantage of others. In due course increased economic activity can be expected to generate additional employment; rising incomes will support greater expenditures on goods and services. Over time those who thought themselves disadvantaged will also benefit from the new economic environment that has been created and the new opportunities for employment and enterprise that it offers. Meanwhile, government initiatives can be taken to alleviate the interim situation. These can include measures to generate employment such as food for work. Part of the tax revenues from successful commercial processing can be used to promote the improvement of traditional processing for low income consumers, such as the local drying of fish in parts of India. Translation into nutritional improvement of rising incomes and employment will also benefit from an appropriate health and nutrition oriented social infrastructure.

ENVIRONMENTAL ISSUES

As elsewhere, processing plants in the developing countries can have a negative impact on the environment in which they operate. Sites were selected at one time for an abattoir and tannery in Maseru, Lesotho on the basis of convenient access from the main road into the city and proximity to the river. No account was taken of the probable growth of the city in those directions. Abattoirs, tanneries and other plants likely to emit unpleasant odours and effluents should be located away and down wind from residential areas.

Provided they are considered sufficiently in advance the amenity implications of a particular location for a plant, noise, traffic congestion, effluent disposal and other adverse aspects, can generally be taken into account without serious increase in cost. A small outlay on tree planting and protection during the early stages of growth can improve the local environment. Such provisions can be part of any official financing support. Environmental hazards are often corrected by measures to reduce waste and improve product handling, as in the conversion into fertilizer of blood and animal waste from abattoirs. The pelleting of cassava in Thailand brought environmental benefits. Formerly dust from the bulk handling of dried cassava chips spread over a wide area south of Bangkok.

The expansion of particular lines of agricultural production to serve processing outlets can also have environmental implications. In the early 1980s there was concern that the rapid extension of cassava growing in some drier areas of Thailand would bring soil degradation and erosion. Government intervention might be needed in such situations to secure the adoption by farmers of protective rotations. Similarly, over fishing might call for the establishment and monitoring of catching quotas.

CONCLUSIONS

The material and human resources available to most government development institutions are limited. This points to a policy of maintaining a favourable policy climate for processing by independent specialized enterprises. Monitoring by a qualified government unit should aim at eliminating obstacles to efficient operation, foreseeing the scope for timely assistance, and intervening to check abuses and minimize undesirable side effects only where necessary.

The overall strategy would be to maintain a stable but stimulative frame for enterprise. This should maximize productive efficiency and achieve a sustained agro industrial development. With its forward and backward linkages the development contribution of agricultural and fish processing can be very substantial. Its limits, as pointed out by Goldsmith A.R. (1985 The private sector and rural development: can agribusiness help the small farmer? World Development 13 (10/11)) are set by the dimensions of the production sectors in developing countries from which supplies are needed to match effective demand for processing.

ISSUES FOR DISCUSSION

1. In what areas of agriculture and fishing have processing enterprises provided a significant stimulus for production? Review the different ways in which this stimulus has been transmitted and assess their relative effectiveness.

2. Assess the advantages and disadvantages of the production contract. How has it worked in your country?

3. In what ways have specific processing enterprises contributed to technology awareness and management capacity in your country? How could their contribution be extended?

4. Specify the institutions in your country offering training supportive of food and agricultural processing, and the courses they offer. Evaluate the content of these programmes in relation to evident needs.

5. List the agricultural and fish processing enterprises earning significant amounts of foreign exchange for your country. What are the trends and the main factors influencing them? How do you see the position 10 years hence?

6. What measures have been taken and programmes established to promote export earnings in your country? How far do these help existing and prospective agricultural and fish processing enterprises?

7. What direct and indirect taxes are paid by agricultural and fish processing enterprises in your country? What concessions do they receive in their initial years? What services are normally provided for them by central and local government?

8. Make your own social assessment of the benefits and hardships accruing from some processing projects in your country. If, in some cases, the balance seems negative what could be done to redress it?

9. Examine some processing enterprises in your country for their environmental implications: if corrective measures seem needed propose a time span over which they could be taken.

FURTHER READING

Abbott, J.C., (1987) Agricultural marketing enterprises for the developing world, Cambridge, Cambridge University Press.

Asian Productivity Organisation (1978) Asian food processing industries, Tokyo, APO 4-14 Akasaka, 8-chome Minato-ku.
Beckford, G., (1972) Persistent poverty: under development in the plantation economies of the third world, New York, Oxford University Press.
Burbach, R. and P. Flynn, (1980) Agribusiness in the Americas, New York, Monthly Review Press.
Harper, M. and R. Kavura, (1982) The private entrepreneur and rural development, Rome, FAO.
Holst, C., (1987) Innovation and entrepreneurship for development, Journal of Development Planning (18).
Kent, G., (1987) Fish, food and hunger, Boulder, Co., Westview Press.
Lal, D., (1978) Men or machines, Geneva, ILO.
Lappe, F.M. and J. Collins, (1978) Food first: beyond the myth of scarcity, Boston, Houghton Miflin.
Mikoti, B., (1985) Agriculture and employment in developing countries; strategies for effective rural development, Boulder, Co., Westview Press.
Morrissy, I.D., (1974) Agricultural modernization through production contracting, New York, Praeger.
Roy, E.P., (1972) Contract farming and economic integration, Danville, Interstate Publishing Co.
Skorov, G.E., (1978) Science, technology and economic growth in developing countries, London, Pergamon Press.
Tavis, L.A., ed. (1982) Multinational managers and poverty in the third world, Notre Dame, Ind., University of Notre Dame Press.
Uma Lele (1975) The design of rural development: lessons from Africa, Baltimore, Johns Hopkins Univ.
UNIDO (1978) Industrialisation and rural development, New York.
Von Braun, J., et al. (1987) Non traditional export crops in traditional smallholder agriculture: effects on production, consumption and nutrition in Guatemala, Washington, International Food Policy Research Institute.
Williams, S. and R. Karen, (1985) Agribusiness and the small farmer: a dynamic partnership for development, Boulder, Co., Westview Press.

5 Planning and management

"Good management sees the opportunity and what must be done to grasp it." This is how the head of a major fruit processing enterprise in South East Asia summed up the foundation of its success. Preconditions for the establishment of an agricultural processing enterprise are:

a) there must be a market for the processed product;
b) there must be capital available to develop the organisation;
c) there must be a reliable source of the raw material.

These three requirements must be met. Capital is essential to finance plant and equipment and purchase supplies. Without a market a processing operation cannot continue. Without a steady supply of suitable raw material it will not be able to hold its market. If there are good prospects of meeting these requirements one can go on to consider the processing technique that would be most effective under the prevailing conditions, the nature and strength of the competition that will have to faced and what location would be most advantageous for the plant. Availability of reliable and competent staff to run it is another major issue in planning a new operation.

In this chapter we shall examine further these considerations drawing on the enterprise case studies and other relevant experience. We shall then go on to discuss management of a processing enterprise as a business, accounting and other controls that will be helpful in its operation and, finally, criteria of performance.

PLANNING OPERATIONS

Identify the market

The essential starting point for any processing enterprise is a clear idea of the market to be served and a plan of action for serving it. Reconnaissance visits to potential distributors and other buyers to ascertain their requirements, coupled with studies of available information on trends in sales of the produce envisaged would be a minimum preparation.

The larger the scale of operation envisaged the more important is it that systematic research on the markets open to it be undertaken as a guide to policy. Market research is needed initially to plan a new operation: it may be needed again at intervals to plan an additional processing line or to maintain and develop an existing market share.

A market research programme for processed agricultural products on accessible domestic or export markets should cover:

1. supply and price trends overall and for different qualities and forms of presentation including sizes of cans, by year and by season, also covering possible substitutes;
2. prospective consumer demand; identification

of target consumers for the product, the market share that could be obtained and ways of increasing it;
3. sales methods and agencies and their respective costs;
4. official regulations and market preferences relating to plant inspection and sanitary requirements, product quality standards, containers, labelling, etc.;
5. quota regulations and levies affecting volume, cost, and timing of sales in particular markets.

Inquiry along these lines is designed to answer the following questions: What is the size of the market? What are the predominant consumer preferences? What is the scope for market segmentation i.e. for separating consumer groups likely to pay different prices according to the quality and presentation of a product? What are potential bases for such segmentation - quality features, packaging, branding, form of presentation, type of retail outlet? Is demand for the fresh product seasonally limited? How would it respond to measures to extend product availability by processing? Is demand likely to grow? what are the determining factors? Is demand seasonal or year round? What financing would be required to cover shipping costs, delays in obtaining settlement, sales promotion, etc.?

Export markets. Trends in production, imports and consumption of processed products on potential foreign markets can be identified by an analysis of official statistics. However, assessment of commercial demand for specific processed consumer items such as canned fruit and vegetables requires detailed accurate information on quality, packaging, timing of deliveries, transport, storage, credit, sales promotion, pricing, trade rules and procedures and access to markets, in particular tariff and non tariff barriers. Potential importers are often the quickest and most accurate source of such information. Such research should end up with recommendations as to which consumers could be supplied with advantage, i.e. higher, medium or lower income consumers, institutions, caterers, their location, preferences and supply channels. An onion drying plant, for instance, was set up in Sudan on the assumption that it would use a red variety of onion. Later, it found that it could not sell its products because consumers preferred the white variety. Consumer preference surveys are used increasingly as a guide in the presentation and processing of produce, or to reveal reasons why sales of long-established items are falling off. Specialized agencies will undertake them for a fee related to the size of sample considered necessary for reliability, amount of travel involved etc.

The organisation of the market is also vitally important. For example, if the wholesale market is managed by a few large houses instead of by many brokers and jobbers a single supplier may not be able to secure a small share of that market. Determination by market research of the buyer structure can provide valuable guidance for policy on sales methods, packaging and presentation. Individual product market studies with addresses of sales agents and distributors are available from the International Trade Centre, Geneva, the Tropical Development and Research Institute, London, and from lcal chambers of commerce.

These studies, however, rarely provide the information required as a basis for producing and marketing a specific processed product. Likewise commissioned market analyses may still be too general if they use national or annual data or assume that related products are more homogeneous than they actually are. The following telegram concerning a marketing consultant for a pre-investment project in Southern Europe shows what is involved:

> Preliminary consideration food marketing problem indicates following specific detailed information required from agencies Western Germany, England. A. Basic reaction distributors to prospects market entry new firms. B. Comparative advantages canned and frozen products.

C. Specifications quality levels and packs each product, form. D. Methods and costs introductory activities. E. Minimum quantities assortments required for establishing, maintaining position. F. Prospects private label versus contract packing including potentialities for combining the two. Consultant valuable this stage only if able supply details from intimate personal knowledge specific European firms.

Domestic markets. The content and coverage of domestic market research and the techniques used need not differ substantially from those applicable to developed country markets. What is critically important, however, is to avoid carrying over into domestic market studies assumptions regarding consumer attitudes, behaviour and response that are often implicit in the teaching of developed country institutions.

The assessment of demand for processed food products on national markets requires knowledge in depth of the consumer groups to be supplied and of their tastes, preferences and purchasing power. The majority of developing country consumers are unable to pay for elaborately processed and packaged foods. To them canned fruit, vegetables and juices appear excessively priced in comparison with the same product fresh, dried or pickled. Often the can costs more than the raw material. The convenience factor, which counts so much with consumers in high income societies, is much less significant in developing countries. This often means that the demand for processed fruit and vegetables is largely seasonal. A study of market opportunities for tomato paste in Ghana showed that the main demand was in the rainy season when fresh tomatoes were not available. Ketchup processors in East Africa found that their mass consumers liked a deep red colour. Higher income customers preferred a lighter red shading towards brown.

In some of the countries classed as developing incomes have been rising quite rapidly as will be evident from Table 5.1. In others with a low average income there are substantial pockets of consumers

Table 5.1 Income per person: averages for developing countries, 1984 (dollars per year)

Country	Income	Country	Income
Bangladesh	130	Malawi	180
Benin	270	Malaysia	1,960
Bolivia	540	Mali	140
Botswana	960	Mauritania	450
Brazil	1,720	Mauritius	1,090
Burkina Faso	160	Mexico	2,040
Burma	180	Nepal	160
Burundi	220	Nicaragua	860
Cameroun	300	Niger	190
Central African Rep.	260	Nigeria	730
China	310	Pakistan	300
Colombia	1,370	Panama	1,980
Congo Republic	1,140	Papua New Guinea	710
Costa Rica	1,190	Paraguay	1,240
Dominican Rep.	970	Peru	1,000
Ecuador	1,150	Philippines	660
El Salvador	710	Rwanda	280
Ethiopia	110	Senegal	380
Ghana	350	Sierra Leone	310
Guatemala	1,160	Somalia	260
Guinea	330	Sri Lanka	360
Haiti	320	Sudan	360
Honduras	700	Tanzania	210
India	260	Thailand	800
Indonesia	540	Togo	250
Ivory Coast	610	Trinidad & Tobago	7,150
Jamaica	1,150	Uganda	230
Kenya	310	Venezuela	3,410
Liberia	470	Zaire	140
Madagascar	260	Zambia	470

Source: World Bank (1986) World development report, Washington

with incomes that are sharply higher. While the average for India was estimated at $260 per person in 1984, five percent of households, and that could total 40 million people, had an annual income over $1,000. With living costs relatively low, this leaves considerable disposable income to spend on food that has been processed. However, adoption of new food expenditure patterns often lags considerably behind a rise in income. Previous tastes are retained. Servants are available to prepare food in the house; they may also do the buying. This an experienced firm like Hindustan Lever found to its cost when it misjudged the market for freeze dried peas. A project for their preparation had to be abandoned after a plant had been constructed at Ghaziabad and farmers trained to produce the raw material to exacting standards.

Realistic identification of the market is essential. If no clear market can be defined, it is best to abandon the project. Where there is doubt it is wise to start on a small scale and adapt the processed product to meet market requirements as they become manifest. Private enterprise and transnational processing managers can shift their operations to match market demand insofar as the available raw materials permit. Managers of farmer cooperatives and parastatals may be subject to specific constraints. A commitment to buy for processing all the produce that is offered to them may be one. This can mean that secondary outlets must be found for products that do not match the requirements of the most favourable markets or exceed the needs of known large buyers who would sign contracts for specific quantities. It is still open to these managers, however, through the prices they pay to growers and provision of technical and financial assistance, to direct production towards market requirements.

Raw material supply

Determination of the source of raw materials for processing is vital. Even a slight deficiency in supply can undermine the viability of a project serving competitive markets. Raw materials of good, uniform quality are needed from sources close to the plant to keep down transport costs. It should also be feasible through staggered planting and harvesting of the crop to maintain delivery schedules that permit capacity operation of the plant over extended seasons. The most common and perhaps the most difficult problems faced by agro industries pertain to the supply of their raw materials. In addition to cost, three other aspects of raw material supply are of vital concern: quantity, quality, and seasonality.

Quantity. In most locations harvests of agricultural raw materials are substantially affected by the weather. The supply of produce to processors also depends on the prices paid to producers and the producers' perception at planting time of how easily they will be able to sell their produce. Many farmers who produce their own food can choose to withhold supplies from the market if prices are not attractive. A trap which befalls many planners of new processing enterprises is to base their operations on a seasonal surplus of supplies on existing markets for fresh produce. This led farmers to complain of low prices. The entry of a processing enterprise would mop up this surplus. When processing begins the quantities of produce available at the low prices observed previously are often much less than was expected. The entry onto the market of the processing enterprise has raised prices against it.

That surpluses and low prices have occurred in some years also does not mean that production will be expanded substantially when a processing enterprise is established to provide a sure outlet. A tomato processing plant was established at Balkatonga in Northern Ghana under the expectation that, with irrigation water available, production would expand to meet its needs. This did not occur; the plant never operated at a viable percentage of capacity. Corn Products Corporation went to Pakistan expecting to buy its maize at a reserve price on the open market.

It never obtained enough and had to make elaborate arrangements to generate additional supplies.

If the commodity in question is new to producers, special support measures to raise their yields and reduce their risks, both technological and market related, may be needed to induce enough production. Most of the fruit, vegetable and sugar cases demonstrate the need for assistance to growers of raw material for processing. That provided by Siam Pineapple proved insufficient: it had to resort to direct production.

In deciding whether a guaranteed price will attract sufficient quantities by expanding the area planted, or whether changes in production technology are needed, the processor will have to look at past responses to prices, at yields on farms that are well managed and at yields in other areas with similar conditions. The probability of success with each approach should be assessed along with the costs of implementation. Furthermore where the product is new to producers, or the plant is located away from traditional trade networks, the enterprise may have to make its own arrangements for buying and hauling raw materials, including investment in its own field roads and transport vehicles. Since this can be a costly undertaking, it must be considered carefully at the time an agricultural or fish processing project is proposed.

Complaints by a group of sugar cane growers in Kenya that they were not served by the existing plants led some men skilled in setting up sugar processing equipment to start their own processing operation. In the light of the complaints by existing producers they thought it would not be necessary to undertake cane production directly. However, the farmers could not agree on a group arrangement to transport their cane to the plant. The company lacked transport of its own, so individual farmers had to obtain their own transport which involved higher costs. Many of the farmers insisted on payment in cash on delivery: otherwise they would sell to pre existing artisanal jaggery processors. While the company had a sure outlet for sugar in the Kenya National Trading Corporation, it could not obtain sufficient cane to cover its costs. Its financiers urged it to establish a nucleus estate, but by this time its losses were such that it could not find the means to do so.

The implications of government pricing policies need careful assessment. If a minimum price is guaranteed for a product envisaged for processing and allows an acceptable operating margin for the processor, then this is favourable. There remains the issue, is there the infrastructure and finance needed for implementation of the price guarantee? If not, the supplies expected may not materialize. The implications of price guarantees for other products must also be taken into account. Output of sunflower in Turkey, an important raw material for Unilever Is, fell when the government revised upwards its guaranteed price for wheat; this was the traditional crop of the farmers who grew the sunflower.

Quality. It is important that the raw material available for processing be suitable. Traditional crop varieties may have the loyalty of local consumers and certain appealing characteristics, but they may not fit the technology used in processing. This issue came up in an FAO project to assist orange juice processing in Libya. The oranges available were of a range of varieties with different juice quality, pit content and skin types. The Malaysian pineapple canning industry found itself handicapped by the variety 'Singapore Spanish' grown there. This pineapple was relatively small, had deep "eyes" and somewhat fibrous flesh. The average recovery for processing was 16 to 20 percent as compared with 24 percent or more for varieties grown in other countries. Quality on arrival at the plant is also affected by the handling involved in bringing produce from small farms over poor roads. To ensure that supplies have the right characteristics for storage, processing, or sale in non traditional or distant markets, the enterprise or the government may need to undertake variety trials, the development of new varieties, the supply of improved inputs, and technical assistance to growers. Cadbury, in India, had to persist with the development of the Forastero cocoa variety when the government had settled on another less suited to its processing operation. Such efforts must be supported with a price structure that places a premium on the desired characteristics such as low moisture content and uniformity in grain, or particular sizes and shapes in fruits and vegetables. Unilever Is applied a consistent set of deductions as an incentive to oilseed crushers to achieve established standards of colour and freedom from fatty acid.

Seasonality. Enterprises processing crops that are harvested once a year face acute problems in managing their operation economically. Often the plant can be kept running only for a few months per year. During this period it may be overloaded. At the peak of the season supplies waiting to be processed may deteriorate. Double and triple shift operations to cope with deliveries at such periods risk that the plant will break down and should be watched carefully. A restricted season of operation raises costs both through low plant utilization leading to high overhead costs and through higher direct costs, overtime and extra charges for transport, etc., when the plant is operating. This problem also occurs, to a lesser degree with animal products. Some Kenya Co-operative Creameries plants operated only for six months of the year due to lack of milk during the dry season. In considering plant size, a balance must be struck between the benefits from economies of scale that most industries could realize during periods of good supply, and the costs of unutilized capacity during the off season. Over optimism with respect to raw material supply and markets has led to widespread under utilization of capacity in agro industrial investments.

Ways of reducing the impact of seasonal variations in raw material supply include:

a) Installing capacity to process a range of raw materials, such as different fruits and vegetables, as they become available. The cost of extra product specific equipment, which is usually required at the first stages of processing, may well be offset by the reduction in fixed costs of capital per unit of output that results from reduced idle time, or by the increase in sales revenues from using the plant more intensively during its period of operation. The Basotho asparagus canning enterprise added beans in sauce for the local market and extra fine beans and canned peaches for export, to cut down idle time. Pineapples are followed by citrus at the Libby plant in Swaziland.

b) Providing assistance to growers to extend the harvest period through changes in the crop varieties grown and in cultivation practices, and by making better inputs available. With the aid of sprinkler irrigation, suppliers to the Findus freezing plant in central Italy grew five crops of spinach and other vegetables per year.

c) Adjusting producer prices according to a pre announced schedule, to encourage deliveries early and late in the season.

d) Drawing raw materials from different areas, with different climates or crop cycles, in rotation.

Figure 5.1 sets out the processing seasons for fruit, vegetables and fish in Tunisia. It provides some pointers for complementary use of processing facilities, provided transport and other considerations are favourable. It also indicates peak seasons of demand for processor finance, container supply and product marketing services.

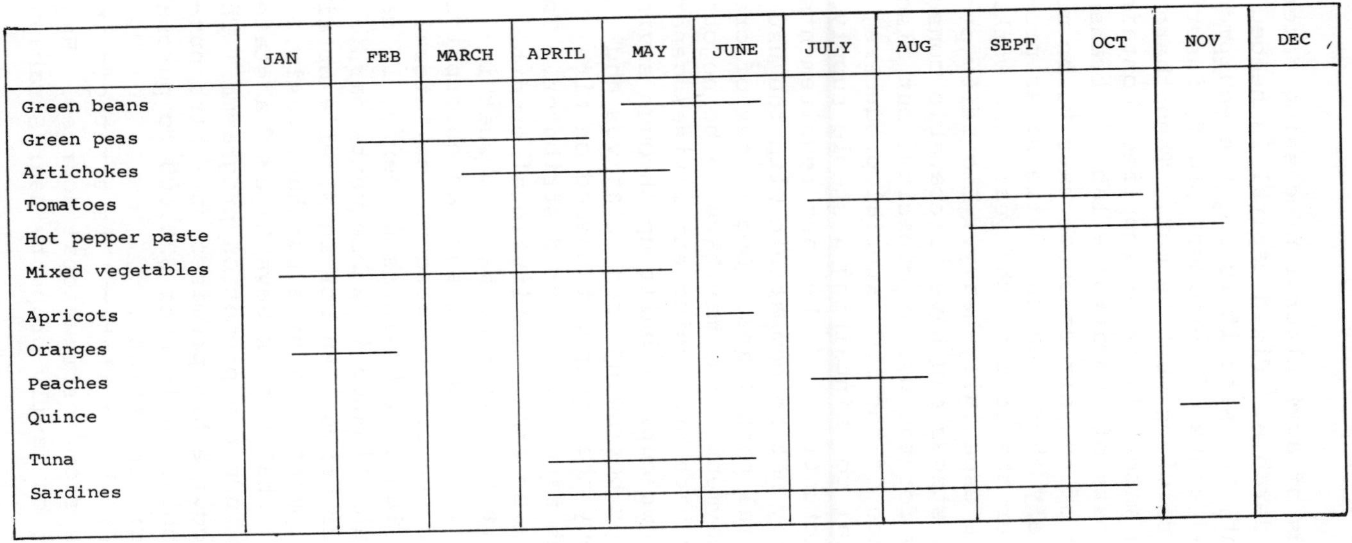

	JAN	FEB	MARCH	APRIL	MAY	JUNE	JULY	AUG	SEPT	OCT	NOV	DEC
Green beans												
Green peas												
Artichokes												
Tomatoes												
Hot pepper paste												
Mixed vegetables												
Apricots												
Oranges												
Peaches												
Quince												
Tuna												
Sardines												

Figure 5.1 Processing seasons in Tunisia

Sources of finance

The capital needs of agricultural processing enterprises are considerable. Fixed capital is needed to pay for the plant, the site it is on, the equipment needed. Overhead costs of permanent staff, insurance, local taxes and services must be met. Then there are the costs of raw material, other supplies, containers and the direct costs of operation - labour, transport, etc. Since most of the direct costs have to be met during a relatively short season, while sales proceeds come in over the rest of the year, the working capital burden is especially heavy. Liquidity problems at critical stages can have irreparable consequences not only for the use of capacity, but also for the confidence of the raw material producers, which is essential to maintaining a stable supply. The seasonal fluctuations in the cash requirements of a processor may not be compatible with the usual terms of commercial bank loans. Open lines of credit and flexible repayment terms may have to be negotiated.

The smaller private processing enterprises generally begin with some own capital built up through savings and add to this by borrowing from relatives and friends. They buy the equipment needed on hire purchase from the manufacturer or distributor. Then they go to a bank or other institutional credit source for a short term loan to make up what is needed to begin operations. If such an enterprise is successful and can provide adequate collateral such as a house, land, or business property, it can expect further bank finance in subsequent years, possibly on easier terms. Enebor in Nigeria obtained rice milling equipment on a hire purchase basis. He rented the site. Chalam took advantage of a state government small industry promotion programme. This enabled him to acquire his premises on a hire purchase basis and obtain a loan of $20,000 to purchase equipment and begin operations.

Ability to offer some collateral is strategic in obtaining finance from institutional sources. Haji Mansur started up on family property near Surabaya.

When he obtained a firm contract to supply the government stabilisation agency at pre announced prices, he was able to borrow $80,000 from a bank to mechanize his mill. According to his financial statement for 1979 (Table 2.2) he also had an overdraft of $100,000. With this he could lay out $75,000 in advances to farmers to assure the mill of supplies. Similarly, Hanapi and Sons in Malaysia obtained a long term loan from the agricultural bank to construct their mill and meet other initial costs. In addition, it had an overdraft of $400,000 in 1984 to buy paddy from farmers.

Bringing in partners who can contribute some initial capital is a common way for a private enterprise to start up on a commercial scale. Jamhuri Tannery began with four shareholders then increased them in number to 10. To help finance its proposed expansion it would bring in additional shareholders with $8,000 of new equity. A capital loan of $24,000 was to be obtained from the Tanzania Rural Development Bank. The National Bank of Commerce would allow three monthly overdraft facilities to finance the much larger purchases of raw hides and skins involved.

A transnational joint venture facilitates the financing of equipment and essential supplies, skilled management and technical know how that have to be imported. Typically these are brought in by the foreign partner as his capital contribution. The machinery may have been used before, as in the case of the Corn Products Corporation plant in Kenya; but the technical management that comes with it can be invaluable. The capitalization of Unilever Is was a similar arrangement with the local partner financing initial outlays in local currency and current operations.

Cooperative enterprises face an inherent disadvantage in mobilizing finance for processing. If members are to contribute capital equally and they are to include small farmers, the total sums raised can only be small. Capital for fixed investments in processing and to finance initial operations must be available from the government or a cooperative bank set up by it for this purpose. The Bunyoro Cooperative Union raised only $2,500 by direct subscription. Acquiring its plants with government assistance, it was able, however, to pay off the loans. In 12 years it had built up its capitalization to $200,000 by ploughing back its annual surpluses into bonus shares.

Under Operation Flood dairy cooperatives in India were helped to acquire equipment, etc. by allocations of dairy products under aid programmes. The proceeds of sale of these products provided finance for the cooperatives.

In many countries governments are providing substantial assistance in the financing of agricultural processing enterprises. This can include direct provision of capital to set up a state processing enterprise, and guarantee of short term loans from banks, as with the Dairy Industries Enterprise of Bolivia and the Botswana Meat Commission. More generally they help make capital available on favourable terms to private and cooperative enterprise. This can be done by instruction to government owned banks - viz. Bank Pertanian finance for Hanapi & Sons and Straits Fish Meal in Malaysia. In accordance with government policy the Bank of Bangkok has financed processing for export at an interest charge of 7 1/2 percent, well below the market rate. The Technocrat programme under which Chalam obtained finance was an Indian state government project.

Cash flow problems are common where processors are constrained to finance a long term investment with medium term loans. Recommended for a programme to encourage the installation of improved olive presses in Tunisia was a two year grace period followed by repayment over the next six years.

While the main focus of the aforegoing has been on how finance can be found for processing enterprises, it is also important that such finance be used to the best advantages. Processing management and financial management are closely related. The proper coordination of stocks and operations requires an intimate knowledge of the process in question. The flow of materials is particularly important because most of the raw materials and intermediate goods used are perishable. Good coordination can minimize a plant's requirements for working capital; conversely excessive or unbalanced inventories, and unsatisfactory purchasing and processing schedules, can easily cause liquidity problems. Managers of agro industrial enterprises need enough autonomy to make their own operational and financial decisions in a timely manner.

Private and cooperative processors are generally sparing in capital outlays on their plants, recognizing that unnecessary expenditures come from their own pocket. The projects that are criticized as over elaborate or grandiose tend to be those set up for governments or under external aid programmes. Concern to make a favourable impression or to meet specific aid expenditure targets over rides economy.

Appropriate technology

This text does not purport to cover the technology of processing: so only a few general points will be made in a context of economic viability.

Traditional processing technologies have generally evolved over long periods to suit local conditions. Where their continued application results in evident loss of product yield and quality, upgrading these techniques can bring significant benefits. Examples are the use of improved milling machinery to raise the yield of rice from paddy, the pelleting of cassava to avoid mould growth and physical (and environmental) losses in its handling as flour, the use of solvents in addition to mechanical crushing of oilseeds to obtain higher oil yields, manipulation of tobacco to eliminate impurities, use of hygienic facilities and refrigeration in the slaughtering of animals and handling of meat, milk, fish and crustacea.

New technologies can be applied to advantage where they open the way to remunerative markets and there is capacity locally to service and maintain them. Straits Fish Meal acquired new mechanical driers manufactured in Thailand. Allana had no difficulty in servicing its quick freezers in Bombay. Ice making and freezing plants operate successfully, with international assistance, in Nouadhibou, Mauritania.

Most of the more advanced technologies can also be transferred to developing countries. Patents and processing secrets are few in the processing of conventional foods and the major non food raw materials. There could be difficulties with sophisticated products such as high fructose sweeteners, protein concentrates, enzymes, dehydrated yeast or specialized foods. These processes involve some patents and require a high order of technical knowledge. Nevertheless, the obstacle is commercial rather than technical, generally insufficient demand from the domestic market.

The most critical issue in the choice of a technology is its compatibility with the environment in which it is to operate. To meet some difficulties firms can take compensatory measures but at a price. Thus, technicians can be trained to maintain and repair complicated imported equipment, but the time and efficiency lost during their training must be weighed. Deficiencies in utilities and other infrastructure can be offset by investment in back up generators and other equipment, but the costs of this may be prohibitive. Problems of this order are recurrent in countries with chronic shortages of foreign exchange and where imports are subject to bureaucratic controls. Reliance on sophisticated technology may also reduce a firm's flexibility and thereby limit its ability to cope with seasonal flucutations in its inputs or changes in its markets. Certain features of the environment will be beyond a firm's control, such as difficulties in importing replacement parts and interruptions in the supply of packing materials. An otherwise attractive project in the Mtwara region of Tanzania based on the use of oilseed crushing mills imported from India foundered over difficulties in obtaining spare parts and maintaining the machinery. Shortages of packing materials and of cans have led to vastly reduced utilization of dairy plants and canning factories in Africa. Where ample labour is available at low cost the sequence of processes should be examined carefully to see where labour can be substituted to advantage. The handling and sorting of raw materials and packaging of processed products can often be done more cheaply manually than with mechanization.

Assessing the competition

In the 1950s the meat processing and marketing transnational Liebig set up an abattoir at Kosti in Sudan. It was conveniently located for transport of processed products to export markets and within droving distance of large traditional herds. The price level in the area for live animals was one that offered an attractive operating margin for the processor. The plant never received a significant flow of stock. After a few years it was dismantled and its equipment sold. The sponsors had neglected their competition. The Kosti enterprise threatened the marketing empire of the then 'cattle king' of Sudan. He had the family ties, traditional allegiances, and financial resources to draw away from the plant the bulk of the raw material supply it had foreseen. He marketed the animals live through the traditional channels offering higher prices than the Kosti plant could pay.

Both on export and domestic markets potential competition must be identified and appraised. If it is effective and has power the options are three:

1. To join forces; Liebig could have invited the 'cattle king' to become a partner in a joint venture enterprise.
2. To abandon the project.
3. To find ways of matching the competition technically and price wise.

Korean and Taiwan exporters of processed produce to Western Europe accepted the third option, the challenge of supplying high quality at lower prices, and succeeded.

> The discipline of producing to the demands of overseas markets can be a great benefit to a developing economy. It isn't noticed much by the economists, but it is likely of more value than the technology and other fruits of modern production.

Lee Kuan Yew, Prime Minister of Singapore, caught the headlines reiterating this view. Such a discipline is particularly relevant for success in selling processed products on export markets.

For products converted into a form that permits keeping over substantial periods of time there is little advantage to be derived either from seasonal differences in harvesting, or proximity to the market. On cost there can be strong competition from other developing countries. Dehydration of vegetables for export to soup and processed food manufacturers has attracted much attention among African countries. The dried product is packed in airtight plastic bags inside metal drums and is light in weight. Packing material and transport costs are low in relation to value of the product. However, ease of shipment over long distances opens up the area of competition. To succeed on international markets a tight integration of production, processing and marketing is needed, as is meticulous attention to quality. This calls for a high level of organisational ability and technical know how. To keep down overhead costs the plant should also operate for most of the year.

On domestic markets in developing countries there is some tendency to under rate the competitiveness of existing small scale rural industries. Their equipment can seem outdated and their products of low quality. However, they may operate at very low cost with their equipment already amortized and labour paid only when required to work. They can rely on local suppliers and local customers and can easily adjust their output to suit current conditions.

Large plants, on the other hand, must often assemble their supplies and distribute their products over a wide radius, involving substantial transport costs and higher buying and selling charges. Sometimes, large processing enterprises have to buy their supplies at prices fixed officially, while their small scale competitors can avoid such controls and buy more cheaply. This was a problem for palm oil processors in Eastern Nigeria for many years.

Small plants with traditional equipment and minimal labour forces also do not face the cost of paying their employees all year round. Machine operators in highly technical processing operations need considerable training and constitute a valuable asset to the plant. This type of skilled labour has to be maintained throughout the year, whether the plant is operating or not. Furthermore, the products of traditional industry, although seemingly of low quality by developed country standards, may be preferred by indigenous consumers because of their familiar style and flavour. Thus, vegetable oil from modern mills has not been able to compete on some local West African markets with oil from the traditional small-scale processors. With a fast turnover it could be sold unrefined, so keeping a flavour preferred by indigenous consumers. In Turkey, however, Unilever Is competed with domestic enterprises in producing and marketing new products that replaced the traditional alternatives because they were much cheaper. Unilever Is succeeded because its quality was consistent and its overhead costs low.

Plant location

Recounting the sequence of decision making over the Del Monte pineapple production and processing operation in the Philippines, a spokesman stressed that availability of the raw material was first priority. This called for a favourable climate, adequate land area with suitable soils, means of transport to the processing plant, and access to labour. Location of a processing plant was further determined by:

a) access to transport for materials to the project and finished products to the market;
b) adequate power, water and waste disposal facilities.

Following a provisional decision in favour of the Philippines, Del Monte made test plantings throughout the country. It was some 10 years later that a site was selected, a plateau of 600 metres altitude with a cool climate and evenly distributed rainfall. It was then sparsely populated so an adequate land area could be obtained. The plantation site was not too far from the coast where the processing plant was located. A deep water port was established for receiving the materials required for the project, and, later, for the shipment of the processed products to overseas markets.

As the population was sparse, labour had to be imported from the coast and nearby islands. At the processing site, a more densely populated area, adequate labour was available. A suitable source of water was found at the processing site, but power and other utilities had to be installed. Major civil works, roads, housing and buildings had to be undertaken both at the plantation and the cannery site.

In general, plants using raw materials that are bulky relative to their value and fairly easily stored, are best located conveniently for raw material supply, e.g. grain mills, cotton ginneries. Indeed, with his mill located in Sidoarjo town, Haji Mansur felt disadvantaged. The transport of paddy from distant production areas had become increasingly expensive. Additionally, he was incurring opportunity costs retaining for his processing operation a site with a much higher potential value for urban development.

Decisions on the location of livestock slaughtering plants call for analysis in depth. Formerly there was little choice. Animals had to be slaughtered near to the point of consumption; otherwise the meat would deteriorate before it could be sold. With refrigeration meat can now be transported long distances. Access to this technical alternative, and its wide use in developed countries, has led various aid agencies to recommend it for developing countries also. They have been struck by the long journey animals had to make on foot, the loss of weight involved, and the poor conditions under which often the animals were handled both en route and on arrival at their destinations.

The transport of meat is generally more economic where road or rail communications are good, where a smooth sequence of handling can be organised, and where the market accepts chilled meat. Under developing country conditions, however, construction of abattoirs in producing areas can often be premature. It calls for investment in chilling facilities, refrigerated truck or air transport from processing point to the market, and refrigerated display cabinets at retail outlets. Keeping this equipment in good running order is likely to be more difficult than in the countries where it is in general use. The risk of deterioration or spoilage as a result of transport delays can be serious.

The nature of the market must also be taken into account. Some developing country consumers have a strong preference for fresh meat, so chilled meat will be discounted. The savings in weight in transporting dressed carcasses instead of live animals are substantial; but in poorer countries most of the fifth quarter goes for human consumption along with the meat. If it cannot be offered fresh at the main consuming centres, it may have to be disposed of at much lower prices; so a significant part of the potential market value of the animal is not realized. Thus location of abattoirs in producing areas distant from their main markets calls for a specific study of costs and returns taking these various aspects into account. This is illustrated in Table 5.2. Between Gao and Accra slaughter in the producing area and air transport of meat were not viable. For Chad, in contrast, they were; movement of animals live to

Table 5.2 Comparative costs and returns from export of live cattle and processed carcasses from Gao, Mali to Accra/Kumasi, Ghana

Live animal	$	Carcass	$
Sale price Kumasi	163	Sale price Accra	189
Less Ghana import duty	23	Hide and by-products sold locally	3
Net sale value	140	Less Ghana import duty	49
		Net sale value	143
Costs		**Costs**	
Price paid to grazier	69	Price paid to grazier	69
Buying fee	3	Slaughter and cold storage	10
Vaccination	1	Transport to airport	2
Export tax	3	Air freight	51
Trekking fee	8	Weight loss	2
Transport in Ghana	11	Sales cost in Accra	1
Transit duty Burkina Faso	2	Total costs	135
Ghana veterinary fee	1	Net return	8
Weight loss 5%	6		
Other costs in Ghana	6		
Total costs	110		
Net return	30		

Source: Fenn, M. (1977), Marketing livestock and meat 2nd edit., Rome, FAO.

its best markets was hardly feasible and the duties levied on meat there were less protective.

Tanneries follow the slaughterhouses which provide their raw material, with easy access to water an additional consideration. Fish cleaning, packing and freezing, and fishmeal plants, locate at the ports where their supplies are landed. Increasingly, as the range of fishing extends, cleaning and freezing take place on the boat: the processed product is landed ready for marketing.

A frequent reason why new processing plants in the developing countries are unable to compete successfully on the markets intended is wrong location. This decision is often tipped by political considerations, ranging from government policies for regional diversification to the personal interests of influential politicians. This can mean poor access to road or rail systems, poor water supply and facilities for waste disposal or lack of a reliable supply of electric power. In such situations the additional investments needed such as provision of standby electricity generators should be explicitly considered in project proposals and financial assistance sought in compensation. Potential project sponsors should try to gauge the adequacy of the facilities likely to be available by looking at the location and experience of existing firms. Still more serious are handicaps such as high costs of transport to the plant of the raw materials it uses and of imported supplies such as containers and fuel, and uncertain distribution services for the processed product. These handicaps may endure for the life of the enterprise. A practical recommendation in planning a new operation is to minimize visibility until its location has been settled. Publicity at the planning stage risks a perverse intervention.

Selection and management of staff

Many small processing enterprises begin as a family operation or one of working partners. They are familiar with the supply, processing and marketing procedures involved through previous experience. Often it is a father with working sons, brothers, or partners who are already friends of mutual confidence, that constitute the operational nucleus. Such small private enterprises then take on additional labour as and when they need it. A shift to higher technology in such an enterprise comes when a son, for example, has had an opportunity to acquire specialized knowledge through a specific course of training.

Staffing becomes a major issue when a large scale enterprise is set up from the start and the operation envisaged is new to the country concerned. Management and technical staff may have to be brought in from elsewhere. They can be hired individually on the basis of response to advertisements in appropriate journals, be provided by a foreign partner in the case of transnational joint ventures, or come from an aid agency by arrangement through the national government. Expatriate staff are expensive. From the outset arrangements should be made for the training of nationals to replace them. This training must be in depth, both as regards understanding of the equipment and procedures involved, and in the number of personnel trained in order to allow for replacements. Often new processing activities have operated successfully for some time after the departure of expatriate personnel, but then became run down as standards of maintenance relaxed, and in face of operating problems that had not been foreseen. Ingenuity in adapting and improvising to meet changing situations is an essential requirement of a processing manager.

As a general principle it is important that the terms of employment of paid staff provide a clear incentive for performance. Salaries and prospects should be attractive so that employees are keen to hold on to their jobs: otherwise they should be paid by results. Strategic personnel, on whom the success of the operation depends, should receive a substantial annual bonus related to plant output and profitability.

All down the line an employee will work better if:

a) he is properly trained in the job and knows what to do;
b) the desirable results of whatever he is asked to do have been defined;

c) guidelines or limits in terms of policy, expenditure, and time have been established;
d) he is left alone to do the job;
e) he knows he can go to his boss at any time for guidance or support when he reaches an impasse;
f) he knows he will not be berated if things do not work out exactly as the boss wanted;
g) he is immediately praised for the things he does well.

A staffing problem faced by the larger processing operations in developing countries, particularly those owned cooperatively, or by parastatals, is pressure to employ too much labour. While increasing employment is an important political and economic objective, especially in remote or depressed areas, inflated labour costs can have a devastating effect on financial performance. Similarly, inability to release unproductive personnel because of employment conditions set by governments or by powerful labour unions, can be a continuing weight on both plant performance and enterprise profitability. These problems can be alleviated if they are anticipated at the time projects are developed. Careful financial planning can provide a basis for negotiation with the bodies concerned in maximizing employment.

A high rate of staff turnover is also a handicap to efficient operation. Cooperative and parastatal enterprises following government wage scales often lose their best people to private enterprise paying higher rates. Processing enterprises located away from the major commercial centres see staff they have trained move away to more attractive living conditions. Providing housing and other on site amenities for workers can reduce labour turnover. There are good grounds for cost sharing between government and a processing enterprise, both in the provision of these facilities and in staff training.

In conclusion, one may consider the advice on how to succeed with a new processing operation, presented with tongue in cheek by an African tomato paste manufacturer at an FAO training centre:

1. Wait until some over capitalized foreign aid or government financed plant has got into difficulties; then take it over at a low market valuation with exemptions from taxes.
2. Invite the president, minister of commerce or other important local personalities to become shareholders. This will protect it from political pressure and, with any luck, from competition.
3. Instead of employing a specialized food chemist, plant engineer, accountant, purchasing officer and sales manager, as often recommended by consultants and aid agencies, the operator should do all these jobs himself together with his brother.
4. Avoid having to observe government salary scales and fixed working hours. Instead pay everybody double for the work they do, and retain flexibility.
5. When the operating season is over spend the time looking for new markets.

MANAGING A PROCESSING BUSINESS

Pricing

The scope for initiative by a processing manager in setting prices for the products and services he sells is determined by the competition he faces on the markets open to him. His position can range from that of irregular seller of small lots on markets served by many other enterprises to that of a monopolist who can release his product onto a protected market at a rate designed to maintain a target price.

The small irregular seller tries to ascertain what is the going price for his product by studying available information sources and approaching potential buyers. If his product is good and his supplies limited he can try asking a little more than the going price. If sales at this price are too slow, then he will have to reduce it to move his stocks. Preferably, this should be done on terms that also reduce his costs. Thus discounts in price for increasing quantities purchased under one order can save money on transport, handling and payment collection costs. Discounts on sales for cash reduce collection costs and also bank financing charges.

The pricing of processing as a service relates logically to the quality of service offered. With new equipment a rice miller can obtain high yields of whole rice grains and separate bran from husk. For this he should be able to charge more than an old mill with a much lower out turn. The degree of competition is still an important factor. Situated in a newly developed production area that was acquiring a certain 'cachet' for quality, Enebor in Nigeria earned much more from milling other peoples' rice than growing his own. In Taiwan, by contrast, with efficient rice mills in every village, the charge for milling paddy is barely more than the value of the bran.

The monopoly processor-marketer sets his sales price to maximize profits. He can supplement his personal knowledge of the market with studies in depth to estimate the elasticity of consumer demand with respect to price. This information would tell him how much more consumers would buy if the price were reduced and how far sales would drop if the price were raised. Using estimates based on these studies he can then decide what quantity to put on the market in order to maximize returns to his enterprise and the prices at which this would be achieved. Thus, while Internor, the agency of Interbras distributing frozen lobster tails on the United States market, had not a full monopoly, it had considerable market weight because it controlled most supplies originating from Brazil. Stocks were held in deep freeze warehouses. Buyers for restaurants were inclined to accept the price requested because they knew there would be no lower priced supplies of lobsters from the same major source.

Between the extremes of competition and monopoly there is a range of intermediate positions. Enterprises offering quantities too small to affect the market in which they sell must generally follow it. They can, however, try to sell part of their supply to regular customers at a premium, on the basis of familiarity to consumers or some other attribute of quality or convenience. This is the position of farmers and some family scale processors who sell part of their output retail, with the rest going to a wholesaler at whatever price it will fetch. A firm with a new attractive product can decide to skim off the price insensitive segment of the market. To open up outlets for a new processor of fairly standard products a low price may be asked initially. This would be raised gradually as the firm's position became established.

Where prices of standard quality processed products are reported regularly from recognized distribution centres, these are widely used as a reference in pricing. Competition between brokers and handlers will have determined a margin between the price at this market and at the location of a particular processor that reflects transport, handling and transaction costs. Sales proceed on the basis of the central market price less this margin (made up of these costs). Parastatal exporters of standard products such as cocoa and coffee tend to follow the established international markets. To enable them to clear their warehouses for incoming supplies, and reduce their bank borrowings, they sell a large part of their total supply at the current spot price. The rest is sold forward over the months when they can expect conveniently to ship it. For such products the spot and forward prices on the London and New York markets reflect the best expert opinion.

Where a processor provides credit to buyers to facilitate sales he will try to cover the cost in his sales terms. Either the cost is charged directly or the price is raised to meet it.

Purchasing

Processing begins with the production, or procurement of raw material for treatment. In planning production or purchasing the processing manager will take into account:

1. Characteristic of the product. Does he know it sufficiently well to assess its quality and value? Can he transport, handle, process and pack it for future sale without incurring substantial waste and deterioration? Can it be processed into a form that will appeal to consumers?
2. Ability to finance. Does he have sufficient own capital available or will he have to go to a bank for credit? Will he be able to get enough credit and at what cost?
3. Price likely to be obtained on sale. Is this a sure price? If it depends on free market determination is the current trend up or down?
4. Prospective profit. Most processing marketing managers incorporate this into a target operating margin which they use in deciding when and at what price to buy.

This operating margin is made up of:

1. Direct costs: payments for labour, handling, transport, sorting, cleaning, additional materials needed in processing, containers, fuel, sales, short term finance.
2. Overhead costs: office expenses, salaries and social contributions for continuing staff, capital loan charges, depreciation of plant, equipment and facilities. These are estimated per unit quantity of product on the basis of recent records.
3. Remuneration for management and risk: this is the net income of the operator or his enterprise.

For convenience, processing enterprises adopt for their operating margin a standard mark up. So a manager will buy for processing what he thinks he can handle, that he can finance, and that he believes he can sell with an adequate mark up. This will be his standard analysis. If he foresees some risk over the resale price he may reduce the quantity purchased; or he may decide to carry the risk in order to please his regular suppliers and to be sure of satisfying regular buyers.

Break even point. Critical for all enterprises with a substantial investment in equipment and facilities is the break even point. This is the volume of throughput at which fixed costs are covered by income from sales after allowing for the costs directly associated with those sales. Fixed costs are interest on capital loans, depreciation of equipment, insurance, rent, salaries of permanent staff, i.e. costs that must be paid irrespective of the volume of business handled. Variable costs are payments for raw material and supplies, for labour and services required only when the plant is in operation and interest on the financing of raw material purchases and stocks of product. The break even point is where sales income is equal to total costs i.e.

$$\text{break even output} = \frac{\text{fixed costs}}{\text{price} - \text{variable costs per unit}}$$

If for a given enterprise fixed costs are \$30,000 per year, the price received for its product is \$700 per ton and variable costs are \$300 per ton, then

$$\text{break even output} = \frac{30{,}000}{700 - 300} = 75 \text{ tons}$$

Optimum throughout. Determination of the level of throughput that is most profitable will also take into account the impact of increasing and decreasing quantity on the supply price of the raw material and the sales price of the processed product. As the quantity of product to be sold increases the net return obtained will tend to decline. Existing markets became saturated. To sell more can mean accepting lower prices or spending more on transport and sales promotion. Conversely, with a smaller quantity to sell higher prices might be obtained with lower sales expenses. In the same way the cost of raw material would tend to increase with the volume of throughput. So a likely relationship between fixed costs, variable costs and volume of throughput would be that shown in Table 5.3.

Table 5.3 Illustrative fixed and variable cost and profit relationships with volume of throughput

	Throughput in tons		
	50	100	150
	(.	$	)
Costs			
Fixed	30,000	30,000	30,000
Variable	15,000	30,000	45,000
Total	45,000	60,000	75,000
Income	50x850 = 42,500	100x700 = 70,000	150x450 = 67,500
Profit	(2,500)	10,000	7,500

With a throughput of only 50 tons the margin between the buying price of the raw material and selling price of the product is highest, $850 per ton; but the weight of overhead costs puts the enterprise in deficit. Around 100 tons of throughput the operation maximizes net income. With 150 tons higher unit costs of raw material and/or lower unit sales returns for the processed product bring down the profit margin between them to $450 per ton. Though overhead costs have been spread over a larger volume net income is declining.

For many processing operations it is convenient to secure supplies in advance via contracts with producers. See the Kagoma, Rifhan and Tabasco cases and Jamaica Broilers. This enables the processor to specify variety, quality, maturity and other standards for his raw material, and time of delivery. He can then operate his plant more efficiently and be sure of satisfying his customers' requirements.

Sample purchasing contracts as used by Findus for the purchase from small farmers in central Italy of green beans for freezing, and by Rifhan for maize in Pakistan are presented in Appendix 1.

Timing purchases of raw material to match the operating capacity of the plant is one of the main considerations behind the production/supply contract. It is important that supplies be available to keep the plant operating. It is also important for the processor to avoid excessive arrivals of raw material likely to deteriorate and involve payment disputes. Unilever Is never established direct purchasing links with farmers for its raw material because of the high seasonal concentration of farmers' sales. Everything had to be bought in a three months period for which a team of buyers would have to be maintained and cash and storage provided. It advanced funds, however, to its intermediary buyers to be sure of supplies.

Most processors incorporate price incentives for quality into their purchasing. To test sugar content in cane, acidity in fruit and similar quality characteristics may require laboratory analysis. Facilities for this must be provided and part of the price withheld until quality has been determined. To avoid discouraging producers such a payment system must be developed in mutual confidence; the initial payments should be large enough to satisfy producers' immediate needs for cash. Cooperative processors, in particular, may delay payments to members for substantial periods until their sales returns come in. The Kenya Cooperative Creameries received milk from farmers on credit. Payment was made after 45 days.

Distribution and sales

Effective marketing is essential if a processing enterprise is to succeed. The owner of a small private operation is likely to undertake sales directly. He will draw on his knowledge of his products, of the market and his experience of human attitudes in bargaining.

Establishment of a sales department headed by a sales manager becomes advisable as an enterprise increases in size and its operations become more complex. Selection of a suitable person for this key post is crucial. He should have experience of selling in the markets envisaged: organising ability and imagination will also be important. He should be open to new ideas and on the look out for new opportunities.

At this stage the processing enterprise must consider the distribution channel that is best suited to its requirements. The alternatives include:

a) contacting brokers in a position to arrange sales on domestic and export markets in return for a commission;
b) using existing distribution channels established by one or more wholesalers.
c) sending out its own representatives on regular rounds to arrange sales to retailers;
d) setting up directly owned distribution branches in major markets.

Using brokers to negotiate sales is the simplest course for a new processing enterprise. It limits costs to pre arranged percentage commissions and can open the way to the most remunerative outlets. This is well demonstrated from the fishmeal boom in Peru and more recently by Basotho asparagus sales on the European market.

Established distribution channels, where they exist, offer the advantages of access to a number of outlets without further effort on the part of the processor. This is especially important for perishables since the cost of setting up a reliable cold chain probably cannot be borne by a single enterprise or product line. In some countries the main commercial distribution systems still radiate from the ports of entry, rather than from domestic centres of production. This can mean that the new processor has to develop his own distribution channels. Agents can be appointed to sell at distant locations. They may want exclusive rights in a defined market area as well as an attractive margin or commission if they are to put their full weight into selling for the firm. This has advantages if the volume of throughput is sufficient. At one time Unilever Is in Turkey used 1,600 wholesalers. However, this number was brought down to 700. Distributing a perishable product, the company preferred the direct contact with retailers for its potential influence on the price charged, the sales display used, and above all quality.

Marketing channels and distribution agencies must be selected to maximize sales volume and returns. Margins offered to wholesalers and retailers should be a clear inducement for them to work for an expansion of sales. Where a plant employs its own sales representatives proper training, supervision, and remuneration is essential if they are to promote fully the markets available.

Sales methods should take careful account of buyers' requirements and convenience. Where established quality grading procedures bring out fully the quality features of a processor's product it is best to follow them. The Kenya Tea Development Authority offered tea at daily auctions so that its clients could inspect it before purchase. Widely practiced has been the establishment of directly owned sales agencies in major markets, with sales elsewhere on a bid and offer basis or through brokers. Thus, the Botswana Meat Commission established its own sales agency in London to serve its major market. This agency was financed from Botswana to buy elsewhere, if necessary to maintain supplies to its regular retail outlets, so important was the need for continuity in distribution.

Failure to maintain regular deliveries has various adverse consequences. Buyers will be reluctant to enter into firm contracts and to promote onward distribution. The price obtained will be that for fillers around established products. Opportunities for discounts on transport and sales charges may be lost.

Close supervision of physical distribution is vital where the product in processed form is still perishable. This is well illustrated by the handling of Unilever margarine in Turkey. Before the use of refrigeration became general this called for considerable ingenuity. Unilever Is pioneered the use of evaporative cooling based on local materials. Trays of margarine in transport were covered first with clean waterproof covers then with straw mats that had been soaked with water. If they dried out during a journey the driver had to find water to keep them moist. Drivers were instructed to park vehicles under shade wherever feasible.

Distribution of Cadbury chocolate products in India proceeded through local wholesalers to retailers. Deliveries to wholesalers were made by carriage and freight agents using insulated vans cooled by blocks of ice. Resale prices recommended by the processor and the margins they allowed to the wholesaler and retailer are shown in Table 5.4. These margins, amounting to about 25 percent of the retail price, were for a branded, relatively durable product in wide consumption. They were to be shared between wholesaler and retailer.

Ability to offer short term credit can be very effective in facilitating sales. How far an enterprise should go in this direction depends on its cash flow position and its competition. Many meat wholesalers finance retailers for a few days: thirty days' credit is common for more durable products. In a strong sales position, Cadbury in India could require that its wholesale distributors pay cash for their supplies.

Table 5.4 Wholesale and retail prices of Cadbury cocoa products in India, 1984

	Price to wholesaler	Retail price
	(. . . $ per unit	. . .)
Milk chocolate 80 g.	.47	.64
Fruit and nut bar 40 g.	.27	.36
Eclairs 100 g.	.44	.58
Bournvita 500 g.	1.57	2.00
Cocoa 200 g.	1.00	1.32

Advertising

How best to promote sales and how much to spend on this are continuing issues for the marketing manager of a processing enterprise. It has always been difficult to assess response to advertising; yet few firms have felt they could do without it.

Some promotion is essential to make potential buyers aware of an enterprise and of what it sells. For products that can be branded a much larger outlay is justified. Differentiated by brand promotion from others basically similar, they can be set in a continuing higher priced category. Great here are the returns to scale. A substantial outlay on advertising can be afforded if the volume of sales is large because cost per unit is then quite low. In Turkey, Unilever Is spent about 1.5 percent of its sales income on advertising. Specialized advice should be sought on the relative effectiveness for the purpose of different media. They can vary greatly with local conditions. In Turkey an advisor to Unilever concluded that its advertising budget should be divided in the ratio: press 67 percent, radio 14 percent, shop demonstrations 5 percent, exhibitions 9 percent, display material 4 percent and printed handouts 1 percent. The emphasis on the press was a tribute to an increase in literacy. The radio budget was relatively small partly because Turks without a main electricity supply saved their batteries by turning them off during advertisements. Figure 5.2 outlines an advertisement for Sana margarine. The focus is on women and use in the family. With brand promotion must go rigorous control of quality. Defects in a branded product can hurt greatly its image and future sales.

Figure 5.2

Consumer directed advertising can be very effective and also very expensive. A spot on television in the UK cost in the early 1980s over $1,000 per second. Consumer advertising of 'Best Dressed Chicken' has done very well for Jamaica Broilers in its national market. Mass media advertising can reach wide audiences at frequent intervals. Income groups that would not have television sets may be reached by posters and transistor radio. Messages to consumers should match their ability to grasp them and to follow instructions on usage.

Coordination of processor advertising with other related agencies and services is important. Thus promotion by the processor of a product of recognized nutritional value may be linked with a government sponsored programme to promote better eating habits. Both should coordinate with retailers in the areas covered in having stocks conveniently ready for sale at promotional prices together with explanatory materials and point of sale announcements.

For products sold in bulk and without a brand that carries through to retail sales points, advertising should be directed at trade buyers. Notices can be presented in trade journals and circulars mailed directly to individual distributors, brokers and specialized buyers. The follow up would be visits by sales staff authorized to negotiate sales to meet buyers' requirements.

Packaging

The art of packaging is to combine protection of the product and convenience in handling with a presentation that helps to sell it, at an acceptable cost. In developing country markets where consumers are accustomed to accepting produce loose in their own containers and are primarily concerned to buy at the lowest unit price, expenditure on elaborate packaging would be wasted. It follows that for such markets the high cost of packaging in relation to the value of the product is a continuing constraint on processing.

However pockets of high income consumers constitute profitable if limited markets in most developing countries. Here, attractive packaging and labelling may be needed to stimulate an urge to buy.

Packaging costs can vary greatly from one country to another, depending on the availability of materials, costs of local manufacture, degree of protection provided against competing imports, etc. Thus, in the early 1980s, small metal containers were 30 to 40 percent more expensive in India, Nepal and the Philippines than in some other South East Asian countries. In Pakistan in the late 1970s there was only one company producing cans domestically. It faced a 37.5 percent duty on imports of tin plate, a 10 percent excise duty on cans and a 20 percent sales tax on the wholesale value of finished cans. One response to high costs was to use containers that could be recycled. In the Philippines ketchup bottles were recycled; instant coffee was packed in glass mugs which had a reuse value to the customer. Burlap sacks can be reused three to four times. It must also be recognized that the apparent high cost of containers relative to product in some countries was partly due to the low cost of the raw material and of the labour involved in its preparation. Where the nature of the product permits a choice of containers, the advantages in convenience and appeal to consumers must be weighed carefully against the cost.

From Table 5.5 it can be seen that the carton commonly used in retail sales of rice in the USA cost in 1980 about as much as the rice it held. It was used because of its convenience for self service retailing; its appeal to American consumers also helped to sell more rice. Polythene bags were less expensive, but still added a lot to the price of the rice. For low income countries where retail labour was cheap and people were accustomed to provide their own containers to carry it home, the economy of distribution of rice in sacks was clearly preferable.

Table 5.5 Cost of alternative packagings for rice, USA, 1980

	Carton	Polibag	Sack
Size of package	18 x 28 oz.	12 x 5 lbs.	100 lbs.
	(.	$	.)
Cost per package			
Materials, labour	3.25	2.30	0.40
Overpacking	0.25	0.15	0.05
Miscellaneous	0.10	0.10	0.05
Total	3.60	2.55	0.50
Packaging cost per ton of rice	229.00	85.00	10.00

Transport

The transport concerns of the processing enterprise manager are twofold:

a) to ensure timely deliveries of raw material to the plant without deterioration in transit;
b) to ensure delivery of the processed product to his customers with its quality unimpaired.

Economy of operation calls for a reliable low cost service in each case. Many processors leave the transport of agricultural raw material to the grower. Cattle raisers' cooperatives and private dealers arranged the movement of livestock to the BMC abattoirs in Botswana. In many countries tomatoes are brought to processing plants in animal drawn carts provided by the growers. Intervention by the processor becomes necessary when such informal transport arrangements result in quality losses and delays that impinge on processing operations and add significantly to costs. Dairy plants often deploy bulk tankers along regular pick up routes advising producers how to keep their milk cool while awaiting collection. Refrigerated vehicles may be acquired to ensure that fish are kept cool en route from a boat or port market to a cleaning and freezing plant.

Regular use is the basis of economy in the ownership of transport vehicles. Many cassava drying and chipping factories in Thailand have their own trucks to bring in supplies from growers and deliver the chips to exporters over an extended season. Where the season is short hiring vehicles from specialized transport firms is generally more efficient.

Access to refrigerated vehicles greatly simplifies the delivery of perishables. They are expensive, however, and the risk of losses due to breakdown is high. Until it is clear that their use is justified by the marketing advantage other ways can be used to control product deterioration in transport. For many years Unilever Is in Turkey distributed margarine in ordinary trucks using evaporative cooling. The main supplier of meat from the Sierra of Ecuador to Guayaquil on the coast in the 1980s used an insulated van travelling by night, so avoiding the capital outlay on a refrigerated truck.

Storage

Holding stock is an integral part of many processing operations. It has to be planned carefully. Where the raw material, as in the case of paddy, is harvested only during certain seasons and is fairly durable ability to hold stocks offers a double advantage to the miller:

a) he can buy when the price is normally lowest, just after harvest;
b) supplies are conveniently available to keep his mill operating over a longer season so

reducing his overhead costs per unit quantity milled.

Storage by the miller has the further advantage of economy in handling. Storage elsewhere can mean that the paddy has to be reloaded onto a vehicle for transport to a mill; this could also involve rebagging.

Storage, however, involves considerable costs: viz.

a) interest on the capital represented by the store;
b) interest on the capital represented by the product in store;
c) loss and deterioration of the product in store.

Losses of grain can be brought down to very low levels by drying before storage, use of insecticides, etc. Nevertheless, for storage of grain over eight or nine months to be economic a seasonal price increase of 17 to 20 percent on the initial value is generally needed, depending on the conditions and interest rates applicable. If government pricing policies do not allow such a margin of return on seasonal storage, then it will be uneconomic for the miller to undertake it. This happened in Indonesia during the 1970s when the low ceiling price maintained by the subsidized rice stabilisation agency BULOG led private millers to disinvest in storage.

For many processors of perishable raw materials storage is impracticable. Extension of the operating season is best approached by:

a) offering price incentives for early and late deliveries;
b) drawing from several production zones with different micro climates;
c) promoting the cultivation of varieties ripening earlier or later than the main crop season.

Storage of the processed product is implicit in the whole operation. One of the main bases for processing, for domestic markets in particular, is to convert the fresh product into a form that will be marketable during the season when the fresh product is not available. For processed products that are stable the main costs in storage are the rent or overhead cost of the store and interest on the capital locked up in the product. Storage by the processor has the advantage, if he can carry the finance, of allowing him to manage deliveries to the market so as to maximize the price obtained. If he is obliged by lack of finance to offer large quantities to wholesalers soon after the processing season is over, his bargaining position will be prejudiced and the price could be low.

Storage issues in the sale of processed products on export markets include the following:

a) Stocks should be conveniently placed for loading onto ships when they are available i.e. in port storage if transport from the plant would take too long.
b) If large quantities are likely to arrive on a particular export market at one time, storage for a substantial part should be pre arranged. Otherwise the price is likely to be depressed. This is especially important with perishables such as frozen meat, fish and crustacea.

Taking on a new activity

From time to time most processing managers see opportunities to add to their business. These can be expanding capacity; using an existing processing line to handle complementary products; production of raw material; undertaking additional marketing functions, e.g. direct wholesaling, or retailing; setting up branches or agencies to sell in new domestic or export markets. The decision should be based on the marginal return: i.e. the profits on the new activity after covering all costs directly attributable to it. These net profits can then contribute towards meeting the general overhead costs of the enterprise. Assessment of the profitability of the new activity should not include an allocation of overhead costs since they have to be met whether it is undertaken or not. Use of a work sheet like the following can be helpful in making estimates. The contribution of a new activity may be small or even negative in the first years. By the third or fourth year it should increase substantially if it is well suited to the enterprise and within the capacity of the personnel available.

Scope for additional activities in conjunction with agricultural processing is considerable because of the seasonal concentration of many of the major operations. Thus the Mazenod cannery in Lesotho took on the canning of beans for low income consumers in the region. Although not very profitable in itself, it was at least expected to make a contribution to plant overhead costs that had to be met in any event.

Originally Botswana Meat Commission sold its products in the UK through an exclusive agent. It then saw advantages in having an agency under its full control. The problem facing it and other African meat exporters was that sales to valuable foreign markets might be interrupted to prevent the transmission of exotic strains of foot and mouth disease, or because of drought conditions in the grazing areas. To cushion the impact of such breaks in supply on its customers' loyalty, BMC acquired cold storage and authorized its distribution branch in the UK to buy meat from other sources.

The Jamhuri Tannery case illustrates the process of deciding on an investment in larger scale machinery to expand throughput and improve operating efficiency. It saw ample scope for:

1. turning out more tanned hides and skins for an expanding home market;
2. chrome tanning additional quantities for export; this would add to their value as

Work sheet: Estimating the advantages of taking on a new activity

	Year 1	Year 2	Year 3	Year 4	Year 5
New activity					
Gross additional income	....	....	....	....	....
Less Variable costs					
Labour	....	....	....	....	....
Transport	....	....	....	....	....
Sales	....	....	....	....	....
Sub total	....	....	....	....	....
Less Fixed costs relating to activity	....	....	....	....	....
Financing of new equipment	....	....	....	....	....
Depreciation of new equipment	....	....	....	....	....
Continuing salaries	....	....	....	....	....
Promotional outlays	....	....	....	....	....
Sub total	....	....	....	....	....
Total costs of new activity	....	....	....	....	....
Net contribution to general overheads	....	....	....	....	....

compared with existing exports of hides sun dried or wet salted.

Table 5.6 shows the annual income and expenditure foreseen for Jamhuri Tannery after the expansion had been completed. Using the data in this table Jamhuri Tannery then made simple projections of cash flows over the next 10 years to see if it could carry the capital repayment obligations involved in the expansion. These are shown in Table 5.7.

Cash flow. While a major new investment may prove profitable in the fairly long run, it is important that the enterprise continue to be financially viable in the meantime. It must have money coming in to meet loan repayments when they are due. This is the concept of cash flow. Table 5.7 shows the expected cash flow of Jamhuri Tannery over the 10 years following the expansion. For this purpose loans received and new injections of equity capital are regarded as cash inflow along with income from sales. Repayments of loans and equity capital are cash outflow along with operating and interest costs.

Discounting to present value. Capital outlays today to bring in returns over years to come carry a cost. This is the opportunity cost of not using the capital somewhere else. For a realistic assessment of the benefits to come they must be discounted back to their present value. As a basis for such discounting it is customary to take the real rate of interest in the country concerned. This is the best practicable measure of the value of capital.

Internal rate of return. A criterion of an investment project applied by many financing agencies is the internal rate of return. This is the discount rate at which the net benefits from an investment discounted back to present values just match the present investment outlay. A illustrative discounting table showing the factors to be applied, according to the number of years, for a limited number of rates is provided as

Table 5.6 Projected profit and loss after expansion: Jamhuri Tannery

	Years									
	1	2	3	4	5	6	7	8	9	10
	($ thousands)									
Gross income	517.9	517.9	517.9	517.9	517.9	517.9	517.9	517.9	517.9	517.9
Less										
Operating costs	484.0	484.0	484.0	484.0	484.0	484.0	484.0	484.0	484.0	484.0
Depreciation	5.7	5.7	5.7	5.7	5.7	5.7	5.7	5.7	5.7	5.7
Loan interest	1.8	1.2	0.6	-	-	-	-	-	-	-
Overdraft interest	2.6	-	-	-	-	-	-	-	-	-
Net income before tax	23.8	27.0	27.6	28.2	28.2	28.2	28.2	28.2	28.2	28.2
Corporate tax at 40%	9.5	10.8	11.0	11.3	11.3	11.3	11.3	11.3	11.3	11.3
Surplus after tax	14.3	16.2	16.6	16.9	16.9	16.9	16.9	16.9	16.9	16.9

Table 5.7 Projected simple cash flows after expansion: Jamhuri Tannery

	Years										
	0	1	2	3	4	5	6	7	8	9	10
	(.	. . . . . .	. . . . . .	. . . . .	$ thousands	. . . . . .	. . . . . .	. . . . . .	. . . . .	. . . . .	. . . . .)
Cash inflow											
Loan	24.1	-	-	-	-	-	-	-	-	-	-
Equity capital	8.0	-	-	-	-	-	-	-	-	-	-
Existing assets	27.3	-	-	-	-	-	-	-	-	-	-
Bank overdraft	-	121.0	-	-	-	-	-	-	-	-	-
Sales income		517.9	517.9	517.9	517.9	517.9	517.9	517.9	517.9	517.9	517.9
Total	59.4	638.9	517.9	517.9	517.9	517.9	517.9	517.9	517.9	517.9	517.9
Cash outflow											
Capital investment	59.4	-	-	-	-	-	-	-	-	-	-
Operating costs	-	484.0	484.0	484.0	484.0	484.0	484.0	484.0	484.0	484.0	484.0
Loan interest	-	1.8	1.2	0.6	-	-	-	-	-	-	-
Overdraft interest	-	2.5	-	-	-	-	-	-	-	-	-
Loan repayment	-	8.1	8.1	8.1	-	-	-	-	-	-	-
Overdraft repayment	-	121.0	-	-	-	-	-	-	-	-	-
Capital replacement	-	-	-	-	-	15.1	-	-	-	-	-
Total	59.4	617.4	493.8	492.7	484.0	499.1	484.0	484.0	484.0	484.0	484.0
Net cash flow	-	21.5	24.1	25.2	33.9	18.8	33.9	33.9	33.9	33.9	

Table 5.8 Discounted cash flow after expansion: Jamhuri Tannery

Years	Investment	Income	Operating costs	Benefits	Discount factor at 50%	Present value
	(.	$ thousands	. . . .	)		$ thousands
0	59.4	-	-	-	1.000	(59.4)
1	-	517.9	484.0	33.9	0.667	22.6
2	-	517.9	484.0	33.9	0.444	15.0
3	-	517.9	484.0	33.9	0.296	10.0
4	-	517.9	484.0	33.9	0.198	6.7
5	15.1	517.9	484.0	18.8	0.132	2.5
6	-	517.9	484.0	33.9	0.088	3.0
7	-	517.9	484.0	33.9	0.059	2.0
8	-	517.9	484.0	33.9	0.039	1.3
9	-	517.9	484.0	33.9	0.026	0.9
10	-	517.9	484.0	33.9	0.017	0.6

The present values over 10 years sum to 64.6 - 59.4 = 5.2.

Appendix 2. Table 5.8 sets out the figures for discounting the cash flow following Jamhuri Tannery's investment in expansion. Column 6 shows the discounting factor to be applied at a rate of 50 percent. Column 7 shows the present value of the investment outlay on the top line and below the present value of the benefits through the tenth year following. The present values of the net benefits sum to $64,600 i.e. comfortably more than the present value of the investment outlay of $59,400. The decision on the appropriate discount rate is made after various rates have been tried in similar calculations. That of 50 percent is presented as one that fits with a comfortable margin.

An internal rate of return of 50 percent would be rated highly by most financing agencies. However, the Jamhuri figures may be unduly favourable; no provision appears to be made after the first year for short term working capital to buy raw materials and finance stocks. Presumably this would be forthcoming from a bank in which case the interest charges would add to the operating costs. The risks of changes in operating conditions, in the relative prices of raw materials and finished products, and in management, all call for a generous projected return.

Sensitivity analysis. Where the investment is considerable these risks are best subjected to sensitivity analysis. This assesses the probable effect on the financial viability of an enterprise of changes in market conditions; what would be the impact of a fall in the gross margin on its product, of a decline in the volume of throughput? This is illustrated with hypothetical figures in Table 5.9. A 10 percent drop in gross margin means a 50 percent decline in profits. If the volume handled also goes down by 10 percent the profits fall by 70 percent even after allowing for some reduction in costs. Changes in price or volume are magnified when translated into profits because of the weight of overhead costs.

In considering a new activity full account must be taken of possible negative implications. When a

Table 5.9 Illustrative sensitivity analysis

	Market situation		
	Continuance of present trends	10% drop in margin	10% drop in margin and in volume handled
Gross unit margin on product, $	1.0	0.9	0.9
Quantity handled, tons	1,000	1,000	900
Gross income, $	1,000	900	810
Costs, $	800	800	750
Profits, $	200	100	60
Change in profits, %	-	-50	-70

process for deriving a valuable edible oil from rice bran was developed it attracted great interest among rice millers. They saw the oil as a net gain; the bran residue after treatment could still be sold as a livestock feed ingredient. To justify the investment in the bran processing equipment, however, they would have to increase substantially the volume of paddy going through the mill. They found that this meant buying paddy further afield; the extra outlay on transport costs counterbalanced the additional income from the bran.

Maintaining accounts

With considerable fixed investments in plant and equipment, major requirements of working capital to purchase raw material and other supplies and hold stocks, most processing enterprises need a reliable system of accounts. If he is to remain solvent the owner should know how much he is earning in relation to what he is spending, and how much he owns in relation to what he owes. For these purposes he should have prepared annually an income and expenditure statement setting out clearly whether he has made a profit or a loss over a defined period, and a balance sheet showing his assets and liabilities at a particular date.

Family enterprises will find such accounts useful as a guide to the progress of their business and to possible financial dangers ahead. They will certainly be needed in approaching a bank for credit and making statements of income to tax authorities. Enterprises that are partnerships, joint stock companies, cooperatives and parastatals are generally under obligation to produce such accounts for shareholders, cooperative members and responsible departments of government.

Tables 2.1, 2.3, 2.4, 2.5, 2.7, 2.8, 2.11, 2.12, 2.13, 2.16 and 2.19 present examples of income and expenditure accounts for processing enterprises ranging in size and complexity from family operated rice mills to the Mhlume Sugar Company and the Kenya Tea Development Authority operating 27 tea processing plants. Most of them follow the practice of setting purchases of raw material against income from sales of the processed product, and of milling services in the case of Hanapi and Sons and Enebor. This leaves a gross margin to cover operating costs and profit. The accounts for Hanapi and Sons itemize in some detail the main cost heads involved in rice milling. It will be noted that depreciation was one of the major items. This is because the buildings and equipment of this mill were relatively new. Older plants that had already written off their capital investment could operate at substantially lower cost as a result. Unless, however, an enterprise sets aside reserves for replacement of its equipment when the time comes, the firm would have to seek new capital or go out of business.

Depreciation is calculated to write off the cost of a fixed asset on a straight line basis over the expected useful life of the asset concerned. Common annual rates used are:

a) buildings, storage, cold stores, 10 percent;
b) plant and machinery, 15 to 20 percent;
c) furniture and fittings, 10 to 20 percent;
d) motor vehicles, 15 percent.

Depending on the operating conditions an appropriate rate of depreciation for specialized and delicate equipment may be 25 percent, also for motor vehicles where roads are poor and maintenance uncertain. Actual rates used are specified in the Jamhuri Tannery case study; it follows the practice of charging the maximum rate acceptable to the tax authorities of its country. With corporation income tax at 40 percent in Tanzania a business enterprise did well to take full advantage of such allowances.

Income and expenditure statements show the net profit of an enterprise for a year. This provides a basis for decisions on how much can be paid out as dividend on share capital or as emoluments to directors; how much can be refunded to members of a cooperative as patronage dividend or credited to their share accounts; how much should be assigned to reserves or carried forward to help stabilize the next year's appropriations.

Comparison of the component figures with previous years can help explain why profits have declined or increased. To bring more precision into operating plans managers can be asked to prepare income and expenditure budgets for a coming year. If the actual figures differ markedly from those budgeted an explanation should be sought.

For most business enterprises an important figure is the rate of earnings on the capital invested. If for some time this is lower than that available elsewhere the owners of the capital will be dissatisfied and be inclined to withdraw it. In Table 2.7 Unilever Is show the gross capital employed in the enterprise immediately below net profits after tax. Profits as a percentage of capital employed ranged from nine percent in 1965 to 29 percent in 1983. This latter figure reflected high capacity use of an old plant. A new plant was being established; the additional investment involved was already bringing down the earnings on capital ratio in 1984.

Formal balance sheets for agri/fish processing enterprises are presented in Tables 2.2, 2.6, 2.9, 2.17 and 2.18. Under 'Assets' are shown first fixed assets i.e. buildings, plant, equipment. Current assets follow i.e. stocks of processed product at current valuation, supplies in hand, and payments due to the enterprise. From these are deducted current liabilities viz. money due to be paid out by the enterprise, bank overdraft due to be cleared shortly. This leaves a figure for net current assets. 'Liabilities' include share capital in the case of a joint stock company or a cooperative, long term loans from a bank and reserves held for agreed purposes. The totals of 'Assets' and 'Liabilities' should balance. Figures for the preceding year are usually provided, as in Table 2.17, to facilitate comparison and evaluation of changes between years.

Access to systematic reliable accounts facilitates assessment of the financial status of an enterprise. One useful indicator is its 'net worth', i.e. owner's capital after deduction of indebtedness. The net worth of Straits Fish Meal in 1982 (see Table 2.18) was total assets $541,000 less current liabilities $212,000 i.e. $329,000. Another indicator is the 'current ratio', which is current assets divided by current liabilities. Generally acceptable is a current ratio of 1.6 to 2.0. This is a measure of adequacy of working capital, though trading enterprises can work with lower ratios provided the level of debtors is low, stocks are well controlled, turnover is rapid and prices cover short term financing costs. With current ratios of 2.2 in 1982 and 1.6 in 1983 the Agricultural Bank of Malaysia regarded the liquidity position of Hanapi and Sons as strong. The firm was in a position to settle its current liabilities out of immediately available current assets. Since these usually include amounts due to an enterprise from customers who have purchased its products, the Agricultural Bank also looked at the time it took to collect these payments due. It noted that the average collection period had extended from nine days to 35 days. The explanation that this was part of a sales strategy of giving credit for a month to retailers, was acceptable. In contrast Cadbury in India had decided to require all its customers to pay cash. Presumably under the previous practice of allowing distributors credit the time they took to pay had become excessive.

A strong cash flow position gives a manager scope for initiative and movement. A commonly used indicator of this is the ratio:

$$\frac{\text{net income to shareholders + depreciation allowances + other income net of taxes}}{\text{capital expenditures + changes in inventories + dividend or interest commitments on capital}}$$

A ratio of more than 1.0 indicates a strong cash flow and high capacity to take on new activities and make new investments.

In its evaluation of Straits Fish Meal's request for further financing the Agricultural Bank of Malaysia made the following points:

1. The enterprise had been operating for 15 years and had a good business reputation.
2. The demand for its product was growing. Its customers were well established companies: many had been trading with it

for 15 years.

3. It was well equipped to carry on its business, in terms both of the technology used and the personnel available to direct its affairs.
4. Application of the conventional financial ratios showed that it was a viable concern. In the last four years the current ratio had ranged from 1.28 to 2.44. The acid test was that it could settle its immediate liabilities out of cash, liquid investments and payments due, without having to sell its stocks.
5. Its gross profit margin i.e. gross profits (sales less cost of raw materials and processing but not overheads and taxes) as a percentage of sales had not fallen below 11.5 percent.
6. The average period needed for collection of payments for purchases was consistent with its policy of allowing one to two months credit to customers.
7. Stocks of processed product amounted to about 25 percent of current assets; this was high, but was considered necessary to meet business demands promptly.

The conclusion was to recommend the loan.

Keeping watch on costs

The low operating costs of many family marketing enterprises stem from the direct relationship between outlays on services and the owners' pocket. The more use he can make of family labour and of own resources such as his house, out buildings, and a vehicle, the more he keeps for himself.

Under the pressure of handling a seasonal product, however, a larger scale operator will have to engage whatever services are necessary to carry out the operation. Preferably, such needs should be foreseen in advance and the costs of using alternative suppliers appraised against their quality, reliability and timeliness. When the processing season is over the manager can do his cost accounting, setting out the supplies and services purchased and analysing their impact on profitability. In Table 2.11 major costs incurred by Unilever Is in Turkey are compared as a percentage of sales. Such analyses bring out the relative magnitude of various costs and changes between years, so pointing to those meriting special attention in a subsequent season.

Cost control is just as relevant to parastatal marketing operations as to private firms and cooperatives. Comparison of the current cost of standard operations per ton of produce handled with those of previous years is one of the efficiency measures available to a monopoly enterprise. Thus yield of processed product from the raw material purchased can be compared between years, likewise size and duration of inventories. A study in the USA showed inventory turnover in meat plants averaging 16 times per year. For fruit and vegetable processors it was between four and five times per year.

Total costs of functional activities can be compared in terms of cost per ton of product out turn. This permits a critical assessment of overhead costs. A rise in administration costs per ton of processed product could reflect reduced handlings because of a poor harvest; otherwise it would point to a need to simplify procedures, combine personnel responsibilities and reduce total staff employed. To permit such analyses expenditures on major operating functions, transport, packaging, storage, port handling, should be accounted for separately, not lumped together under such headings as labour, services, etc.

The duration of credits to customers remaining outstanding should also be watched carefully. Collection time is calculated by dividing 'accounts receivable' by the average daily business receipts: up to one month is generally acceptable.

Full use should be made of cash in hand. A large African coffee processing and marketing cooperative was found to have overdrafts carrying interest charges on some of its bank accounts, while in others there was money earning nothing. Consolidation of its accounts, with money not needed until the next crop season put on interest earning deposits, would have reduced its bank charges significantly.

Quality control

Controls on raw material quality at the delivery stage are needed to eliminate deteriorated or otherwise unsuitable produce, also to implement quality incentive pricing to producers. Management of quality in the processed product is still more important. Vital for this are high standards of sanitation with regular sterilization or steam cleaning of all equipment with which food products have contact. For fish and milk handling this is of crucial importance. Maintenance of consumer confidence in the processed product is essential. Up to 1987 much of the marine produce exported from India to Europe had to be defrozen to check for salmonella bacteria; it was then repacked for retail distribution.

Systematic tasting, with laboratory analysis in larger enterprises, is important for consistency of food products. For perishables a maximum shelf life in distribution must be established. In some countries this must be set out clearly on the container; otherwise it can be shown by means of a code known to retailers, distributors and their staff. Unilever Is took back for remelting all margarine that had been two months in distribution. It maintained its own distribution network primarily for quality control.

Where necessary as the basis of a contract, the quality of a processed product can be verified by a qualified independent agency. The Surveillance Générale of Geneva is often employed for this. Its fee is around one percent of the value of the cargo examined. Thus, the final price of cassava chips is determined on their arrival at the port of destination by means of such a survey. Samples are taken from various parts of a consignment. A premium is paid by the buyer, or deductions made according to whether the product is above or below an established quality level.

How far product quality should surpass a level that is acceptable for general consumption in the market foreseen is a marketing issue. Better quality packs may be prepared for more demanding outlets and higher income consumers. A fine balance between quality and price is the goal. Adherence to standards that eliminate much raw material without bringing a more than corresponding return in the price obtained for the product would be wasteful. Quality at the consumer level should be checked periodically by examination of samples from various lots. Competing products should be assessed at the same time for features that consumers might prefer and for relative price.

Insurance

Enterprises engaged in the processing and marketing of agricultural and fish products face a wide spectrum of risk. It ranges from changes in the price of products in response to events fully external to the enterprise to losses of stocks and supplies in storage or transit and misuse of funds by responsible officers.

Protection against losses on forward contracts or stocks due to price changes can be obtained by hedging on a futures market where one is available for the commodity concerned. In many developing countries access to such markets is limited by government policy and foreign exchange controls. However, purchasing agreements with suppliers and sales to customers can be related to prices reported on recognized forward markets. The Karachi market, for example, provided a basis for pricing cotton in Pakistan.

Protection against many other risks can be obtained through commercial insurance. A cooperative can insure against the disappearance of a manager with its funds, by requiring that he take out a bond for e.g. $5,000 or $50,000 according to an estimate of the amounts at risk. A firm undertaking storage of produce for other enterprises can be required to obtain insurance against loss due to fire, its own negligence, etc. Insurance can be obtained against losses of produce shipped on consignment due to delays in transit and mishandling by the transporter or consignee. Where the risks of accidents, illness, etc. sustained by employees or users of processing facilities are substantial, insurance against them should be obtained annually as a matter of course.

Currency risks

Sudden changes in exchange rates can affect dramatically an enterprise engaged in the export of processed products or dependent on some supplies that are imported. If a change in rates is anticipated the operator will do well to budget his outlays on replacements at the rate foreseen for the future.

High rates of inflation are common in many countries. The manager of a processing enterprise must be aware of their implications. Selling on the basis of a 10 percent margin to cover costs and profit will leave him in deficit if the currency has devalued 10 percent in the meantime. To maintain his income in real terms his operating margin must be costs plus target income plus an allowance to cover the expected degree of inflation. A distributing firm interviewed in Brazil in 1983 added a 10 percent margin to its procurement prices to cover its own costs and profit, then a further 50 percent to cover expected inflation. Under inflationary conditions depreciation allowances should also be based on the cost of replacement at current prices, not past costs of purchase or construction.

Use of a computer

Wherever electricity services are reliable, and there is adequate technical support, advantage should be taken of computer technology. Inexpensive equipment is now available to undertake purchasing, sales invoicing and accounting work, and to prepare and address standard letters to suppliers and to potential customers. It can also be used to maintain records of employees, their salaries and social contributions and to make tax and other standard deductions where applicable.

Computers can also be used as a decision support system. They can help a processing manager calculate optimum solutions to problems involving simultaneous equations of input and price variables, and to present alternative supply options. Thus least cost combinations of various feed ingredients to meet a required nutritional formula can be determined. Use of cassava chips for livestock feed in Europe has been favoured by access to this technology. For periodic problem solving computer services can also be hired. For this purpose and for record keeping the effectiveness of the computer depends on the quality of the information fed into it. If this is unreliable or not kept up to date, what comes out will be of equally low value.

ASSESSING PERFORMANCE

Commonly accepted indicators of performance in a processing enterprise are:

- Profitability;
- Return on capital employed;
- Market standing;
- Quality of service;
- Innovativeness;
- Social responsibility.

Profitability

This means that at the end of the year the processing enterprise achieves a net surplus after allowing for all costs. How big that surplus should be will depend on the attitude of the owner. A private operator will expect a profit at least equal to what he could earn in alternative occupations open to him. If he lives in an environment of low wages and high unemployment he may accept a very low net income for lack of an alternative.

The transnational enterprise with other sources of income may be prepared to take a long run view on the profitability of a processing plant in one particular country. A firm with worldwide interests such as Nestlé may put first its public relations profile and be satisfied with maintaining a presence in expectation of future opportunities. More characteristic, however, has been attitudes such as that of Gulf and Western Inc. It sold its sugar holdings in the Dominican Republic in 1984 when low prices and high taxes meant that they contributed less to its earnings than its investments elsewhere.

A farmers' cooperative processing plant may be satisfied with a very small profit. Its primary concern may be to serve its members by providing an outlet for produce surplus to fresh market requirements. For this purpose an income sufficient to cover its costs and to maintain reserves commensurate with its operations would be judged adequate.

For a parastatal processing enterprise profitability may have still lower priority. Stabilising the market for a certain category of producers, earning foreign exchange for the national economy, redistributing income in accordance with concepts of equity can be prior considerations. Unless, however, the government concerned is in a position to subsidize it, a parastatal, like a cooperative, must cover its costs. It will be noted, moreover, that the Botswana Meat Commission while paying out a bonus to suppliers, also built up its reserve for expansion. Reserves also feature strongly in the accounts of the Kenya Tea Development Authority.

Return on capital employed

An adequate return on the capital employed in a processing enterprise is important where the sums involved are substantial and have either been borrowed by private owners or could be used by them in some other income earning investment. The criterion for such investors is that the return is at least as high as could be obtained from other available investment opportunities.

The risk of losing the investment must also be taken into account. If a safe government bond pays five percent interest net of inflation then a private business investment should pay substantially more. For a transnational investing in a country where it faces political, exchange rate and other risks additional to those of a familiar business, a return of 20 percent on external capital invested could be a normal target rate. How far targets are achieved depends of course on commercial conditions and other factors. The return on capital of the 18 largest food processing firms in the USA averaged 10.7 percent in 1981, less inflation. The return on Unilever's investment in Turkey was over 20 percent in 1983 and 1984; in 1965 it was only nine percent. (See Table 2.11)

Returns on capital employed receive less attention in parastatal processing enterprises. Their original capital may have been provided from public sources without obligation to pay interest. Some governments charge interest on fresh capital at a concessional rate and expect the parastatal to obtain bank finance for its purchases and stocks; they help in this by providing a guarantee. While a marketing parastatal may well be justified on service and social grounds, it is appropriate, nevertheless, that it account at market rate for the capital it employs. The advantages it offers may then be set against its full cost.

Market standing or market share

This is another objective measure of a processing

enterprises's performance. If it has a 20 percent share of total sales of a certain product or products as against 12 percent some years ago, then it has performed well in this regard. A declining share would imply the reverse. This measure can be applied to most processing enterprises, including parastatal monopolies where they compete on export markets.

The market standing of a firm is high if its share of the market for its product is at the higher income end, and if it has a reputation for quality products and for reliability in its dealings. It is unwise, however, to rest on such a reputation for long. An economic recession or other adverse shift in the market may find it unable to match more innovative or cost conscious rivals.

Quality of service

This is not easily measured, but is well understood by the clients of a processing enterprise. For the farmer it is continuity of market outlet, full observance of contractual obligations and consistent application of objective criteria in pricing for quality of raw material. Avoidance of undue waiting time for deliveries to the plant is also rated highly, together with prompt payment when due. For the customers of a processor/wholesaler consistency of product quality, reliability in delivering the quantities agreed and willingness to adapt to varying demand requirements, are important quality of service criteria.

Innovativeness

Readiness and capacity to introduce new techniques, adopt new forms of organisation, develop new markets are also important criteria of performance in a processing enterprise. If there is to be progress in a particular sector of agricultural and fish processing there must be innovation. New techniques of supply organisation, processing and marketing are quickly taken up after they have been demonstrated as effective under the prevailing conditions. It is the role of the innovator to identify them and try them out.

Cassava chips became the main foreign exchange earner of Thailand within a decade of the development of new technologies and markets by European transnationals and their local partners. Poultry production and processing enterprises were set up in Lebanon by people who had read of the new technology and associated business organisation, but had no substantial experience of it. Marigold production and processing was unknown in Ecuador until an entrepreneur saw that conditions were favourable for it; likewise the chilling and air transport of meat from Chad.

Social responsibility

Consideration for the welfare of the people with which it is in contact is an intangible, but also very important criterion of a processing enterprise's performance. To a considerable degree there is a coincidence of longer run interests. An enterprise dependent on suitable raw material must treat its suppliers fairly well if it is to expect to have their business the following year. If it goes back on the prices it promised, farmers will look for another market outlet, or go out of that line of production. Too soft an attitude, however, over raw material quality may not be in the interests of the growers concerned. It could mean that the customers of the processor become dissatisfied; its business declines and with it the outlet for the farmers' produce.

While they may vary prices according to quality and to market conditions, most processing enterprises find it advisable to offer their suppliers and also their customers much the same terms, particularly those suppliers and customers whose business constitutes the mainstay of the enterprise. The processors are concerned to avoid a reputation for unfair treatment, favouritism or unreliability.

A cooperative or parastatal may go further in demonstrating equity of pricing and service. It may absorb extra transport costs on produce collected from distant growers, for example. The management should recognize that this involves extra costs, however, and in consequence a service to those producers who are more conveniently placed that is more expensive than necessary. Thus against a high performance rating on social responsibility may have to be set a lower rating for service, and for cost. Striking an appropriate balance between competing criteria of performance is the mark of a good processing manager.

ISSUES FOR DISCUSSION

1. What are the main markets for the processed agricultural products of your country? Review the coverage of the existing market studies for these and potential additional products. What scope do you see for expansion on a) domestic markets, b) export markets?

2. Assess the raw material supply for the lines of processing considered to have good market possibilities. What changes would be needed in the nature, timing and other aspects of production?

3. Construct for your country a chart along the lines of Figure 5.1. What prospects does it offer for complementary use of processing facilities, produce picking and handling crews, etc. Assess the scope for modification of existing seasons by planting early and late maturing varieties, etc.

4. What are the sources of a) fixed capital, b) working capital for processing enterprises in your country? Verify this for specific enterprises where possible and ascertain the constraints on access to such capital.

5. Appraise the concept 'small is beautiful' in relation to the technology used in agricultural processing enterprises in your country.

6. What competition do the agricultural and fish processing enterprises of your country face on a) domestic market, b) export markets? For which would you recommend more protection and which would you open to further competition?

7. What factors would favour location of an agricultural processing enterprise at a port?, in a capital city?, in a main producing area? Review the location of some enterprises in your country in the light of these factors.

8. Prepare, as manager of a new processing enterprise, sales plans for three agricultural or fishery products of your country a) on domestic markets, b) on export markets. Appraise the scope for promotion by various methods.

9. Some processing enterprises commit five percent of sales income to promotion of their products. Comment in this light on the sales policies of the enterprises featured in this text; or those of other enterprises you know.

10. Appraise the packaging used by some of the processing enterprises of your country serving a) export markets, b) your domestic market. What changes would you suggest?

11. Propose an additional activity for one of the enterprises for which the case study includes an income and expenditure account. Construct a set of figures for a work sheet estimating its contribution to the general overhead costs of the enterprise.

12. Refer to the figures provided for Jamhuri Tannery. Comment on their adequacy as a basis for the

expansion envisaged.

13. Construct realistic investment, income and cost data for a new agricultural or fish processing plant in your country. Set this out over a period of 10 years. Ascertain the internal rate of return on the basis of the figures you have used.

14. Examine the income and expenditure statements and balance sheets presented in the case studies. Comment on inconsistencies. Rearrange the accounts to conform with approved practice where appropriate. Substantiate the modifications you propose.

15. Derive from the data provided in the case studies some significant financial ratios. What do they tell you regarding the financial status and efficiency of the enterprise concerned?

16. How far do the processing enterprises you know apply cost control measures of the kind set out in this chapter?

17. Assess the performance of some of the enterprises for which there are case studies; of some other enterprises with which you are familiar.

FURTHER READING

Abbott, J.C. and J.P. Makeham, (1988) 2nd edit., Agricultural economics and marketing in the tropics, London, Longman Group.

Austin, E.J., (1983) Agro-industrial project analysis, Baltimore, Johns Hopkins University Press.

Boydell, T., (1986) Management self development, Geneva, ILO.

Chaston, I., (1983) Marketing in fisheries and acquaculture, Farnham, UK, Fishing News Books Ltd.

Downey, W.D. and S.P. Ericksen, (1987) 2nd edit., Agribusiness management, Hightstown, N.J., McGraw Hill.

FAO Agricultural Services Bulletins on the processing of specific products.
Fisheries Technical Paper No. 1276 (1986) Marketing the products of acquaculture.
Marketing Guides No. 2 (1970) Marketing fruit and vegetables; No. 3 (1977) Marketing livestock and meat; No. 4 (1961) Marketing eggs and poultry; No. 6 (1972) Rice marketing, Rome.

Goldberg, R., (1974) Agribusiness management in the developing countries - Latin America, Cambridge, Mass., Ballinger.

Gupta, V.K. and V.R. Gaikwadi, (1982) A guide to the management of small farmers' cooperatives, Rome, FAO.

Harper, M., (1973) The African trader; how to run a business, Nairobi, East African Publishing House.

Ingle, M.D., N. Berge and M. Mainalton (1984) Microcomputers in development; a manager's guide, West Hartford, Conn., Kermarien Press.

I.T.C. Advisory manuals on export markets for specific products and on export marketing procedures; lists of potential importers, Geneva, ITC/GATT.

Jones, S.F., (1985) Marketing research for agriculture and agribusiness in developing countries; courses, training and literature, London, Tropical Development and Research Institute.

Kotler, P., (1984) 5th edit., Marketing management, analysis and control, Englewood Cliffs, N.J., Prentice Hall.

Price Gittinger, J., (1972) Economic analysis of agricultural projects, Baltimore, Johns Hopkins University Press.

Rao, V.R. and J.E. Cox, (1978) Sales forecasting methods: a survey of recent developments, Cambridge, Mass., Marketing Science Institute.

Schumacher, E.F., (1973) Small is beautiful, New York, Harper and Row.

Timmer, C.P. et al. (1975) The choice of technology in developing countries: some cautionary tales. Harvard Studies in International Affairs No. 32, Cambridge, Mass., Harvard University Centre for International Affairs.

Tropical Development and Research Institute, Advisory manuals on the utilization of specific agricultural products, on processing technology and on markets for products and by-products, London, 127 Clerkenwell Rd. ECIR50B.

UNIDO (1978) Manual for the preparation of industrial feasibility studies, New York, UN.

Appendix 1 Production/processing contracts

Agreement governing purchase of beans for freezing: Findus, central Italy

Dear Sir:

Referring to our verbal understanding, it is agreed that you will sell your entire crop of beans from your land situated in______________________________ ______________________________hectares in area.

The expected date of sowing will be______________

Sales will be subject to the following agreement and conditions:

1. The beans will have a maximum size of____________ millimetres.
2. For every kilogramme of beans meeting these standards we shall pay_____________. Beans not meeting these standards will not be accepted.
3. For produce supplied and accepted we shall make payments at the above price twice a month, by the 20th for produce supplied between the 1st and the 15th and by the 5th of the following month for produce supplied from the 16th.
4. Supplying of beans is subject to our "general conditions," of which a copy is attached and forms an integral part of the agreement.

We await your confirmation of approval by returning a copy of this, and of one of the two attached copies of the "general conditions of supply," both signed on each page.

Yours truly,

FINDUS, INC.

General conditions of supply of beans to Findus, Inc.

1. As a guarantee of quality, the grower commits himself to observe the following cultivation rules:

a) The grower will begin sowing on the precise day indicated by the buyer in order to ensure a consistent supply.
b) The grower will use the exact quantity of seed which will be supplied at the price of______ per kilogramme by the company and will be of the high/low variety.
c) The grower will sow all the above seed and will neither grow beans for, or sell them to, any other enterprise during the same period.
d) The grower will grow the beans on suitable well-fertilized land on sections previously approved by the buyer and according to its instructions.
e) The supplier undertakes to use only those pesticides recommended by the buyer, and at times authorized by it:
   i) up to one month before harvest: metasistox, DDT, parathion, metil-parathion, ethion, trithion and lindane;
   ii) up to 21 days before harvest: rogor;
   iii) up to 15 days before harvest: sevin;
   iv) up to 7 days before harvest; phosdrin.
f) If for weather or other reasons sowing cannot be carried out at the time agreed or should be wholly or partly lost, the buyer may at its discretion furnish seed for another period or abandon this sowing. As each grower understands, each sowing is part of a large programme intended to ensure a regular supply of raw material to the plant. Consequently, the farmer is required to communicate the probable failure of any planting as soon as it is apparent, particularly when it might amount to more than 25 percent of the area grown.
g) The farmer undertakes to keep the area sown weeded according to normal possibilities of good farming. For the use of chemical weed killers he must seek the agreement of the buyer, which cannot, however, guarantee the success of the products it recommends.
h) The farmer undertakes to inform the company in good time of the presence of any insects on the

crop. Application of any chemical preparations should have its agreement. The buyer considers such treatments necessary but cannot assume responsibility for, or guarantee the success of, any one of the various treatments it may propose.

2. The whole harvest will be taken over at the farm, or the nearest place accessible by motor truck, in boxes furnished by the buyer, on which the name of the grower should be clearly indicated.

3. The beans must be supplied in sound condition, and immediately after harvesting, which should be carried out at the correct stage of maturity. Pods damaged by insects, fungus or other causes, as by weather, that are joined together, bent or otherwise not conforming to the correct shape will not be accepted, decision on this being solely by judgement of the buyer. Pods containing beans that are too large will not be accepted on equal terms. The grower will do whatever is possible to avoid quality impairment.

4. When picking up each lot, the buyer will take samples from at least five boxes. These samples will be put in a special sack and examined at the plant. On this basis, the buyer will determine, by its sole judgement, quality and size percentages. These percentages will be the basis for payments according to the agreed conditions and price. The amounts not accepted under Article 3 and following will not be paid for, nor can they be returned because of the impossibility of identification and recuperation once they have been put into the machines at the plant. The grower is invited to be present when his samples are analysed.
As regards dispatch, acceptance, queries, etc., obligations arising from the supply agreement will be considered lot by lot.

5. Any lot with more than 25 percent of beans dirty

or otherwise damaged (not in sound condition, or of irregular size) can be refused by the buyer in its entirety. Lots with less than 25 percent discards but more than 7 percent insect damage will also be refused. A rejected lot will be held at the disposition of the grower for two days after communication of rejection, provided the buyer has storage space available, after which the buyer is free to dispose of it.

6. The grower must proceed with harvesting according to the buyer's instructions. In June, July and August, the hot season, harvest should proceed day by day according to the buyer's instructions. The grower must watch that the produce does not have mixed in it stones or other foreign bodies which might damage machinery or equipment at the plant. The grower declares that he understands that the fresh beans are intended to be deep frozen.

7. The grower is obligated not to hand over or sell any part of the crop to another firm, person or organization, except with the express authorization of the buyer. In cases of inobservance of this, the buyer must be indemnified at the rate of ten times the price agreed under the supply contract.

8. Representatives of Findus, Inc., have the authority to inspect at any moment the growing crop, to check the area and take samples. Their instructions on cultivation must be carried out exactly.

9. Considering that the grower understands that the purchase of fresh beans is tied to their processing for deep freezing, it is expressly agreed that the buyer will be free of any obligations and duties deriving from the agreement, should it be prevented from carrying out the acquisition and processing of the product for reasons beyond its control (burning of the plant, floods, prohibition by law, etc.).

10. Any dispute arising out of this agreement will not be taken to the courts but will be settled by a team of three arbiters composed on a friendly basis, two to be chosen by each of the parties and the third by these two, or failing that, by the President of the Court at Rome at the instance of the more aggrieved party. The arbiters will decide on the basis of equity and will meet in Rome.

The grower, undersigned, declares his express approval of the clauses of the above Articles 1, 3, 4, 5, 7, 8, 9 and 10.

Maize production and procurement contract, Rafhan Maize Products Co., Pakistan

1. The Farmer agrees to plant ________ hectares of maize under the supervision of the agronomist appointed for this purpose by the Company.

2. The Farmer will prepare the land, apply irrigation water and carry out all other practices according to the directions and to the satisfaction of the Company's agronomist.

3. The Farmer will use the seed of maize supplied by the Company at a cost of $7.50 per hectare and apply the fertilizer as prescribed by the agronomist.

4. Planting of maize will be done according to the instructions of the Company's agronomist at the time and for the area prescribed by him. All labour etc., required for planting will be provided by the Farmer.

5. Irrigation of the crop will be the responsibility of the Farmer and he will apply irrigated water according to the instructions of the Company's agronomist.

6. Insect control will be done under the supervision of the Company's agronomist who will prescribe and provide the insecticide required for this purpose. The Farmer will supply the labour and equipment required for the application of the insecticides. The cost of insecticides will be debited to the Farmer's account.

7. The cost price of seed, fertilizer and insecticides supplied by the Company to the Farmer shall be considered an advance payment in respect of the price of maize to be delivered by the Farmer to the Company and shall be adjusted from the total price payable to the Farmer.

8. Hoeing and weeding will be the responsibility of the farmers who will make sure that there is no weed in the field. He will be required to hoe the crop at least three times.

9. Watching the crop to avoid pilferage will be the responsibility of the farmer.

10. Harvesting of the crop will follow the instructions of the Company. Its moisture content should not exceed 28 percent. Moisture content will be tested by the Company's agronomist.

11. The Company will purchase the entire produce of the crop grown under this contract. The price will be $____ per kg. of dry grain (15.5 percent moisture).

12. The Company may purchase the produce on cobs if the Farmer agrees to the following terms and conditions.

    a) Any cob with more than 28 percent moisture in the grain will be rejected.
    b) The Farmer agrees to the calculations of the Company's experts in determining the price of ears based on dry grain weight expected to be received from the cobs. In this case the

total price of the ears will be paid on receipt of delivery by the Company's representative.

OR

The Farmer should agree to keep duplicate sample in sealed bags weighing 10 lbs each. The number of ears in each sample will be counted and recorded. One of the two samples will be retained by the Company while the other will remain with the Farmer. When the ears are dried after approximately 8-10 weeks the samples will be brought together at the company's purchase centre and the grains shelled and dried to 15.5 percent moisture. The grain will be weighed and calculations made accordingly. In this case half of the price will be paid to the Farmers at the time of receiving the delivery of his produce while the balance will be paid when the final calculations are made.

13. In cases where the Company provides the inputs and some how the Farmer is unable to carry out farm operations according to the instructions of the agronomist and it is feared that crop may be adversely affected, the Company reserves the right to get such operations performed at their own expenses and such expenses shall be considered as advance towards purchase price and adjusted from the cost of maize payable to the Farmer by the Company.

14. The produce will be purchased by the Company at the Farmer's place provided the quantity is in truck loads and the place is connected with a good road on which the trucks can ply. In case of smaller quantities the farmer will have to deliver the produce to the nearest purchase centre.

15. In case of any dispute between the parties the matter shall be referred to arbitration under the laws of Pakistan. The Maize Development

Manager of the Company shall be the Sole Arbitrator in such disputes. His decision shall be final and binding upon both the parties.

16. This contract shall be considered for all parties to have been entered at Faisalabad and only the courts at Faisalabad shll have the jurisdiction.

IN WITNESS WHEREOF the parties have set their respective hands at the date and place mentioned above.

Witnesses: Rafhan Maize Products Co. Ltd.

1.________________ ________________________

____

2.________________ ________________________

Appendix 2 Discounting table

Present value of a future lump sum: $$PV = \frac{A}{(1 + R)^n}$$

PV: present value
A: unit amount
R: rate of compound interest per period
n: number of periods

Period	Interest rate, R							
n	4%	5%	6%	8%	10%	12%	14%	15%
1	0.962	0.952	0.943	0.926	0.909	0.893	0.877	0.870
2	0.925	0.907	0.890	0.857	0.826	0.798	0.790	0.756
3	0.889	0.863	0.840	0.794	0.751	0.712	0.675	0.658
4	0.855	0.823	0.792	0.735	0.683	0.654	0.597	0.572
5	0.822	0.784	0.747	0.681	0.621	0.567	0.539	0.497
6	0.790	0.746	0.705	0.630	0.565	0.597	0.454	0.432
7	0.760	0.711	0.665	0.584	0.513	0.432	0.400	0.376
8	0.731	0.677	0.627	0.540	0.467	0.404	0.351	0.327
9	0.703	0.645	0.592	0.500	0.424	0.361	0.308	0.284
10	0.676	0.614	0.558	0.463	0.386	0.322	0.270	0.247
11	0.650	0.585	0.527	0.429	0.351	0.288	0.237	0.215
12	0.625	0.557	0.497	0.397	0.319	0.237	0.208	0.187
13	0.601	0.530	0.469	0.368	0.290	0.229	0.182	0.163
14	0.578	0.505	0.442	0.341	0.263	0.205	0.160	0.142
15	0.555	0.481	0.417	0.315	0.239	0.183	0.140	0.123
16	0.534	0.458	0.394	0.292	0.218	0.163	0.123	0.107
17	0.513	0.436	0.371	0.270	0.198	0.146	0.108	0.093
18	0.494	0.416	0.350	0.250	0.180	0.130	0.095	0.081
19	0.475	0.396	0.331	0.232	0.164	0.116	0.083	0.070
20	0.456	0.377	0.311	0.215	0.149	0.104	0.073	0.061
21	0.439	0.359	0.294	0.199	0.135	0.093	0.064	0.053
22	0.422	0.342	0.278	0.184	0.123	0.083	0.056	0.046
23	0.406	0.327	0.262	0.170	0.112	0.074	0.049	0.040
24	0.390	0.310	0.247	0.158	0.102	0.066	0.043	0.035
25	0.375	0.296	0.233	0.146	0.092	0.059	0.038	0.030

Index